ELECTROMAGNETIC WAVES IN MOVING MAGNETO-PLASMAS

ELECTROMAGNETIC WAVES IN MOVING MAGNETO-PLASMAS

by
BASANT R. CHAWLA
and
HILLEL UNZ

THE UNIVERSITY PRESS OF KANSAS
Lawrence / Manhattan / Wichita

Standard Book Number 7006-0068-X
Library of Congress Catalog Card Number 71-107329
Printed in the United States of America

TO OLIVER HEAVISIDE

(1850-1925)

PREFACE

This book presents a comprehensive treatment of various aspects of electromagnetic wave propagation in infinite and bounded moving plasmas, in the presence of a static magnetic field, for both non-relativistic and relativistic velocities of the plasmas. An extensive bibliography is given at the end, which includes a comprehensive list of books in related areas and a large number of articles related to the present subject. The present book could be used as a text for an advanced graduate-level course or seminar and as a reference to scientists working in ionospheric wave propagation, plasma shock waves, astro-physics, and other fields. A strong background in electromagnetic theory and some background in the special theory of relativity and macroscopic plasma dynamics is assumed.

Many of the contributions in this book are based on the original research work and publications by the authors, published and unpublished research papers by Unz, and unpublished portions of the Ph.D. dissertation by Chawla for the Electrical Engineering Department of the University of Kansas. Various approaches to the problems under consideration are presented and compared. Rather than an end result of past research work by the authors, this book, together with its extensive bibliography, should be considered as a starting point for future research work in this relatively new and exciting field. Some possible applications of the present theory are suggested in the Summary.

We are especially grateful to Dr. C. T. Tai, formerly of the Antenna Laboratory, Ohio State University, Columbus, Ohio, for his valuable comments on some of the original papers by Unz, and to Dr. P. C. Clemmow, of the Cavendish Laboratory, Cambridge University, Cambridge, England, who originally suggested the problem solved in Chapter 6, for his many stimulating discussions with Unz during his stay there. Unz wishes to thank Dr. C. A. Levis, Director, Electro-Science Laboratory, Ohio State University, Columbus, Ohio, and Dr. C. M. Crain, the RAND Corporation, Santa Monica, California, for their generous support of the relevant research work done by him while he was in those two places during the summers of 1961-62 and 1965-66, respectively.

The research work presented in this book was supported in part by the National Science Foundation while Unz was an NSF Postdoctoral Fellow at the Cavendish Laboratory, Department of Physics, Cambridge University, Cambridge, England, during 1963-64, and by NSF Grant No. NSF-GK-465 at the Electrical Engineering Department, University of Kansas, during 1965-68. We wish to thank Dr. L. L. Bailin, Chairman of the Electrical Engineering Department of the University of Kansas, for his continuous encouragement and generous support.

The computations in the present book were done on an IBM 7040 electronic digital computer at the Computation Center of the University of Kansas. This is to thank Mrs. Lynda S. Austin, Mrs. Bobbie J. Huyett, and Mrs. Wilma Cromwell for their competent typing of the manuscript and Mrs. Ruth J. Haug for the indexing of the book.

Basant R. Chawla

Hillel Unz

May, 1969

CONTENTS

LIST OF FIGURES

I

INTRODUCTION

Plasma is considered to be the fourth state of matter; it is neutral as a whole and consists of electrically charged positive and negative particles and neutral molecules. A volume of a partially ionized gas is classed as a plasma if the volume is larger than that of a sphere with a radius equal to the screening distance. The study of plasma dates back to Michael Faraday's observations of the glow discharge in the 1830's. Various other aspects besides the electrical discharge in gasses have been studied since then. They are mainly [Holt and Haskell, 1965]: charged particle aspects of kinetic theory, astronomy in relation to solar plasma, and propagation of radio waves in the ionosphere. The solid state plasma, consisting of electrons in metals and electrons and holes in semiconductors, has been of great interest in recent years. The present book is related to the area of propagation of radio waves in the moving ionosphere and other moving plasma media.

A brief historical account of the ionospheric studies can be found in the book by Ratcliffe [1959]. Budden [1961] and Ginzburg [1963] have given detailed treatments of the problems involved in the ionospheric studies. Their discussion is, however, limited to stationary non-moving plasma. The movements of the ionosphere have been considered by Martyn, Maeda, Ratcliffe, and several other authors and a list of references was given by Unz [1961].

Because of the possible applications to space communications, astrophysical phenomena and laboratory beam systems, the radiation and propagation of electromagnetic waves in drifting moving plasma have drawn the attention of several authors. The propagation of electromagnetic waves in a moving plasma in the presence of a uniform static magnetic field seems to have been first considered by Bailey [1948; 1950]. His interpretation of the transverse waves as amplifying waves was, however, later criticised by Twiss [1951] Several works on the propagation of electromagnetic waves in a uniform, moving, unbounded, infinite magneto-plasma (plasma in a uniform static magnetic field) have appeared since then [Bell and Helliwell, 1960; Scarf, 1961; Unz, 1962, 1965a, 1966a, 1968; Tai, 1965a; Chawla, Rao, and Unz, 1966]. The emphasis in these works was placed on the refractive index equation (characteristic equation), and its possible applications to instabilities and amplification in moving plasmas were indicated.

In contrast to the works on propagation of electromagnetic waves in a moving, unbounded magneto-plasma, the boundary value problem of propagation of electromagnetic waves in a moving, bounded magneto-plasma has received a limited attention only. Landecker [1952] considered the normal incidence of circularly polarized electromagnetic waves on a semi-infinite plasma moving along and transverse to a static magnetic field. Fainberg and Tkalich [1959] generalized the situation to propagation in a magneto-plasma moving through a dielectric medium. Reflection and transmission of electromagnetic waves from a bounded isotropic plasma, in the absence of a static magnetic field, have been considered by Lampert [1956] and Yeh [1966]. The term bounded is used here for a semi-infinite medium or a slab.

Radiation from electromagnetic sources in the presence of a moving medium has received some attention recently. Lee and Papas [1964, 1965] and Tai [1965b, c] considered the electromagnetic radiation in the presence of a moving, simple, isotropic and dispersive media. Lee and Lo [1966] solved for the radiation fields in the presence of a moving, uniaxial, anisotropic media and gave formal solutions for the radiation in a moving magneto-plasma. Using an approach which differs from that of Lee and Lo [1966], Chawla and Unz [1966a] gave a formal solution for the radiation fields from electromagnetic sources in the presence of a moving magneto-plasma.

The present book considers various aspects of electromagnetic wave propagation in unbounded and bounded moving magneto-plasmas. The plasma is assumed to be uniform and in equilibrium. The entire plasma, consisting of electrons, ions, and neutral molecules, is assumed to drift with a uniform velocity through a uniform static magnetic field. In general, the motion of the plasma is assumed to be at an arbitrary angle with respect to the direction of the static magnetic field; the mechanism responsible for this uniform motion is, however, of no consequence here. Small signal, single fluid theory is used here in the sense that the passage of an electromagnetic wave is assumed to cause small perturbations of the electrons from their unperturbed positions, and the equations are linearized. The ions are assumed to be heavy and their perturbation under the influence of the electromagnetic wave is neglected.

In chapter two a brief review of the relativistic transformations for moving media is presented, since they are extensively used in the following chapters. Transformations of the fields, the wave vector-frequency, and the plasma parameters are the important quantities considered. A brief review of the Minkowski and Chu theories for moving media is presented.

In chapter three the non-relativistic magneto-ionic theory for moving, temperate magneto-plasma is discussed, by using the polarization current model and the Chu formulation for moving media [Fano, Chu, and Adler, 1960]. The term "temperate" is used here when the temperature effects are neglected; this is the case when the thermal velocity of the electron is much less than the phase velocity of the electromagnetic wave, but much greater than the induced velocity of the average electron due to the electromagnetic wave in the absence of the magnetic field [Allis, Buchsbaum, and Bers, 1963]. The particular case of the longitudinal propagation case (the whistler mode), where the static magnetic field, the drift velocity of the plasma, and the direction of propagation of the electromagnetic wave are all in the same longitudinal direction, is solved in detail. The refractive index equation (characteristic equation) is found for the general non-relativistic case, where the static magnetic field and the drift velocity of the plasma are in arbitrary directions and the electromagnetic wave is considered to propagate in the $\hat{z}$ direction. An algebraic characteristic equation for the complex refractive index is found, and it is discussed for several special cases. The algebraic characteristic equation of the refractive index is transformed into alternative forms by neglecting the second order terms of the relative drift velocity of the plasma for the non-relativistic solution.

In chapter four the characteristic equation for an oblique electromagnetic wave propagating in an arbitrary direction in a moving, unbounded, temperate magneto-plasma is derived in a determinantal form for the general non-relativistic case. The characteristic equation is obtained by using two models for the plasma, namely, the polarization current model (PCM) used by Unz [1965a] and the convection current model (CCM) used by Bailey [1948] and by Tai [1965a] for the normal incident case. It is shown that the resulting determinantal characteristic equations obtained by using the two models are identical. The general relativistic refractive index characteristic equation for oblique electromagnetic waves in an arbitrarily moving magneto-plasma with arbitrarily directed static magnetic field is derived [Unz, 1968], and particular cases are discussed. The relativistic characteristic equation for the normal incidence case is found from the general characteristic equation, and it is shown to agree with the non-relativistic characteristic equation for small plasma drift velocities.

In chapter five the characteristic refractive index equation for an electromagnetic wave propagating in an arbitrary direction in a moving, warm magneto-plasma is derived. The term "warm" is used when the temperature effects are taken into account and the plasma temperature is assumed to be uniform. The characteristic equations are found, using two methods of derivation. In the first method the non-relativistic equations of plasma dynamics in the laboratory system [Oster, 1960] are used for the derivation of the characteristic equation, which is found in a determinantal form, applicable only for non-relativistic drift velocities of the plasma. In the second method the characteristic equation is first found in the rest frame of reference of the plasma, to which relativistic transformations are applied, in order to obtain the characteristic equation in the laboratory system of coordinates [Chawla and Unz, 1969a]. In this method, the characteristic equation is derived in an algebraic form, applicable as well for relativistic drift velocities of the plasma. The unbounded plasma may have a stationary boundary, where, on one side of which a gas of neutral particles is flowing towards the boundary, and on the other side an ionized gas is flowing away from the boundary; the boundary thus provides a mechanism for ionization. Alternatively, the ionized plasma could be bounded and be drifting along with the same velocity as the moving boundary. Both cases are discussed for the relativistic characteristic refractive index equation. The corresponding longitudinal propagation case is discussed in detail.

In chapter six the reflection from a semi-infinite temperate magneto-plasma is discussed. The incident wave is assumed to be linearly polarized, and it propagates in a direction normal to the boundary. The direction of the motion of the plasma and the direction of the static magnetic field also coincide with the normal to the boundary. The solution for the reflection coefficients is obtained by using two approaches. In the non-relativistic solution [Unz, 1967a], the problem is completely solved in the laboratory system of coordinates. In the relativistic solution [Chawla and Unz, 1967a] the problem is first solved in the rest system of the plasma, to which relativistic transformations are applied in order to obtain the solution in the laboratory system. The results thus obtained by the two distinct approaches for this longitudinal case are found to be identical. Numerical results for the power reflection coefficient are given and discussed [Chawla and Unz, 1968]. Analytical expressions are found for the case of circularly polarized incident electromagnetic waves. The considerations in this chapter show that significant effects are observed when the velocity of the plasma is comparable to the velocity of light in free space.

In chapter seven a linearly-polarized plane electromagnetic wave is assumed to be normally incident on a semi-infinite magneto-plasma moving uniformly through a dielectric medium. Two configurations of the static magnetic field are considered, namely the static magnetic field normal to the boundary and transverse to the boundary. Numerical results for the power reflection coefficient are obtained and discussed for two cases. This case of dielectric media is especially suitable for experimental observations of significant results in bounded plasma moving at non-relativistic velocities, since the Doppler shift depends on the relationship between the velocity of the moving reflecting plasma boundary and the phase velocity of the electromagnetic wave in the stationary dielectric medium.

In chapter eight the reflection coefficients

and the transmission coefficients are considered for the case of a linearly polarized electromagnetic wave normally incident on a temperate magneto-plasma slab, moving longitudinally along a magneto-static field, which is perpendicular to the slab boundaries [Chawla and Unz, 1969b]. This is an extension of the longitudinal, semi-infinite, moving magneto-plasma case to be considered in chapter six. Numerical results for the reflection and the transmission coefficients are obtained and discussed. The formulation for the reflection and the transmission coefficients become simplified for the incidence of circulary polarized electromagnetic wave.

In chapter nine the general problem of oblique incidence of a linearly polarized electromagnetic wave on a semi-infinite temperate plasma, moving uniformly at an arbitrary angle with respect to the static magnetic field, is considered. The reflected waves are obtained for parallel and perpendicular polarizations of the incident wave, and the power reflection coefficient is defined. The results are shown to agree with the results of chapter six for the particular case of normal incidence.

In chapter ten the formal solution of the general problem of oblique incidence of a linearly polarized electromagnetic wave on a warm magneto-plasma slab, moving through a dielectric at an arbitrary angle with respect to the static magnetic field, is given. The results are shown to agree with previous results for particular cases.

In chapter eleven the radiation from electromagnetic sources immersed in an infinite, moving, temperate magneto-plasma is considered. The use of the polarization current model of the plasma in the extension of the Minkowski theory for a moving anisotropic medium, given by Lee and Lo [1966], is briefly reviewed, and a formal solution using the convection current model (Lorentzian viewpoint) is presented.

In chapter twelve the Poynting vector in a moving plasma and moving polarizable media, in general, is discussed. The corresponding Poynting theorems for different formulations are studied, and the Poynting vector is chosen so that the various Poynting theorems lead to the correct identical physical interpretation for the moving polarizable medium.

In chapter thirteen an electric-field wave

equation is derived for the electromagnetic waves in a two-stream magneto-plasma [Chawla and Unz, 1969c]. The two streams of the plasma are assumed to move uniformly with different velocities with respect to the direction of the static magnetic field. The current contributions of the two streams to the Maxwell equations are found by using relativistic transformations. The corresponding refractive index equation will be derived by substituting the exponential space-variation factor of the electromagnetic wave, and by setting the determinant of the matrix of the coefficients of three linear, homogeneous equations in the three unknowns E_1, E_2, and E_3, to be zero. Particular cases are discussed.

In chapter fourteen we give summary, conclusions, and the possible applications of the present book to physical phenomena. An extensive bibliography for the subjects discussed in the present text is given at the end of the book.

The system of units used throughout the present book is the rationalized MKSA system of units. The notation of different terms in the present book follows in most cases the notation used in the book by Budden [1961]. The specific definitions of the terms used are given in the text. In particular, we denote the electron charge by (-e), where e is a positive quantity, and the plasma frequency by ω_p. The gyromagnetic (cyclotron) frequency is defined by $\overline{\omega}_H = e\overline{B}_o/m = e\mu_o\overline{H}_o/m$ and $\overline{Y} = \overline{\omega}_H/\omega = e\overline{B}_o/m\omega = e\mu_o\overline{H}_o/m\omega$; thus in the present text $\overline{Y}$, $\overline{\omega}_H$ is in the same direction as $\overline{H}_o$, $\overline{B}_o$, since e is a positive quantity, unlike the notation used by Budden [1961], where they are in opposite direction to each other. The harmonic time variation factor is taken everywhere in the form $e^{+i\omega t}$, as in Budden [1961].

II

THE RELATIVISTIC TRANSFORMATIONS FOR MOVING MEDIA

2.1 Introduction

By covariance of equations one means the identical form of the equations representing physical laws in two different reference systems. The reference systems with which we shall be concerned here are the inertial frames moving uniformly with respect to each other. An inertial frame is defined as an imaginary coordinate system in which a body with no forces acting on it is unaccelerated.

The covariance of Maxwell equations under a group of orthogonal, linear transformations in space time was first shown by Lorentz [Stratton, 1941]. The transformations were named after him as Lorentz transformations by Poincaré [Sommerfeld, 1964a]. The wide acceptance of Maxwell's electrodynamics rests on its covariance within the Lorentz group. The fundamental invariant of this group is the separation of two world points, described by the four-dimensional line element, which is the separation of two neighboring points in space-time. The result that the Maxwell equations are covariant under Lorentz transformations is of great physical significance. The Doppler effect is a result of this [Papas, 1965]. The Minkowski theory of moving media was based on it [Sommerfeld, 1964a]. It also conforms with the first postulate of the special theory of relativity presented by Einstein in 1905 [Einstein, et. al., 1923]. In fact, the second postulate of Einstein's theory postulates the velocity of light in free space as an important universal constant, which is included in the Maxwell equations.

The Einstein special theory of relativity is based on two postulates which resulted from his analysis of time and space and some ingenious imaginary experiments. The two postulates, which explain the celebrated Michelson and Morley experiment of 1887 [Einstein, et.al., 1923], and many other observations [Panofsky and Phillips, 1962] not explained by other theories, are today considered to be most probable, if not certain. The two fundamental postulates of the special theory of relativity are:

1. The Principle of Relativity: The laws of the states of physical systems are not affected, whether these changes of state be referred to the one or the other of two systems of coordinates in uniform translatory motion.

2. The Principle of the Constancy of the Velocity of Light: The velocity of propagation of light (an electromagnetic disturbance) in free space is the universal constant c which is independent of the reference system.

Using the second postulate, Einstein arrived at the same transformations as those given by Lorentz. He obtained the field transformations by requiring the covariance of the Maxwell equations using the first postulate. By establishing the special theory of relativity, Einstein provided physical insight into the space-time interdependence and the laws of nature, which was lacking in Lorentz' work. The Lorentz transformations can be considered as a consequence of the second postulate of the special theory of relativity. The group of Lorentz transformations, which is extensively used in the present book, is given in the next section.

2.2 The Group of Lorentz Transformations

The group of Lorentz transformations, described in the previous section, concerns transformations in the space-time four-dimensional world. Let us consider two reference systems S and S' which are moving uniformly with respect to each other. Let the coordinates in the unprimed system S and the primed system S' be $x_1 = x$, $x_2 = y$, $x_3 = z$, $x_4 = ict$, and $x_1' = x'$, $x_2' = y'$, $x_3' = z'$, $x_4' = ict'$, respectively, where c is the velocity of light in free space. Then from the second postulate of the special theory of relativity, one has

$$x_j'x_j' = x_jx_j, \; j = 1,2,3,4. \tag{2-1}$$

where Einstein's summation convention on repeated indices is implied and the origins of two systems S and S' coincide at t = t' = 0. The corresponding spatial axes of the two systems are assumed to be parallel to each other.

The above invariance property results from a linear orthogonal transformation of the form:

$$x_j' = a_{jk}x_k; \; x_k = a_{jk}x_j' \tag{2-2}$$

where the coefficients of a_{jk} are to be such that

$$a_{jk}a_{j\ell} = a_{kj}a_{\ell j} = \delta_{k\ell} = \begin{cases} 1 \text{ for } k = \ell \\ 0 \text{ for } k \neq \ell \end{cases} \tag{2-3}$$

The linear transformations (2-2) constitute the complete group of Lorentz transformations. From (2-3) one finds that the determinant $|a_{jk}| = \pm 1$. The positive sign coresponds to the positive transformations and the negative sign corresponds to the negative transformations. To include the identity transformation $x_j' = x_j$, the positive sign is chosen. The positive transformations represent the proper Lorentz transformations, which are the ones to be used in the following chapters.

Under Lorentz transformations a 4-vector A_j, which is a set of four variant scalars, transforms as:

$$A_j' = a_{jk}A_k; \; A_k = a_{jk}A_j' \tag{2-4}$$

It can be easily shown that the scalar product of two 4-vectors is invariant to a rotation in space-time. A 4-tensor A_{jk} of rank 2, which is a set of 4^2 quantities, obeys the following transformation law:

$$A'_{jk} = a_{j\ell}a_{km}A_{\ell m}; \; A_{\ell m} = a_{j\ell}a_{km}A'_{jk} \tag{2-5}$$

where all the indices vary from 1 to 4.

If two inertial frames S and S' have the same orientation, and if their relative velocity is $\overline{v}$, then the coefficients a_{jk} are given by [Papas, 1965]:

$$a_{jk} = \begin{bmatrix} 1+(\gamma-1)\frac{v_x^2}{v^2} & (\gamma-1)\frac{v_x v_y}{v^2} & (\gamma-1)\frac{v_x v_z}{v^2} & i\,\frac{\gamma v_x}{c} \\ (\gamma-1)\frac{v_y v_x}{v^2} & 1+(\gamma-1)\frac{v_y^2}{v^2} & (\gamma-1)\frac{v_y v_z}{v^2} & i\,\frac{\gamma v_y}{c} \\ (\gamma-1)\frac{v_z v_x}{v^2} & (\gamma-1)\frac{v_z v_y}{v^2} & 1+(\gamma-1)\frac{v_z^2}{v^2} & i\,\frac{\gamma v_z}{c} \\ -i\,\gamma\,\frac{v_x}{c} & -i\,\gamma\,\frac{v_y}{c} & -i\,\gamma\,\frac{v_z}{c} & \gamma \end{bmatrix} \tag{2-6}$$

where $\gamma = (1-\beta^2)^{-\frac{1}{2}}$, $\beta = v/c$, $i = \sqrt{-1}$.

The position 4-vector x_j, which can be written as $(\overline{r}, ict)$ transforms as

$$\overline{r}' = \overline{r} - \gamma\overline{v}t + (\gamma-1)\frac{\overline{r}\cdot\overline{v}\,\overline{v}}{v^2} \tag{2-7}$$

$$t' = \gamma\left(t - \frac{\overline{r}\cdot\overline{v}}{c^2}\right) \tag{2-8}$$

Similarly, the current 4-vector $(\overline{J}, ic\rho)$ transforms as

$$\overline{J}' = \overline{J} - \gamma\overline{v}\rho + (\gamma-1)\frac{\overline{J}\cdot\overline{v}\,\overline{v}}{v^2} \tag{2-9}$$

$$\rho' = \gamma\left(\rho - \frac{\overline{J}\cdot\overline{v}}{c^2}\right) \tag{2-10}$$

The field vectors $\overline{E}$, $\overline{B}$, $\overline{D}$, $\overline{H}$ in the set of Maxwell equations

$$\nabla \times \overline{E} = -\frac{\partial\overline{B}}{\partial t}\,;\quad \nabla\cdot\overline{B} = 0 \tag{2-11}$$

$$\nabla \times \overline{H} = \overline{J} + \frac{\partial\overline{D}}{\partial t}\,;\quad \nabla\cdot\overline{D} = \rho \tag{2-12}$$

transform as 4-tensors (antisymmetric six-vectors here) from [Stratton, 1941] $(\bar{B}, -\frac{i}{c}\bar{E})$ to $(\bar{B}', -\frac{i}{c}\bar{E}')$ and from $(\bar{H}, -ic\bar{D})$ to $(\bar{H}', -ic\bar{D}')$, so that [Møller, 1952; Papas, 1965]:

$$\bar{E}' = \gamma(\bar{E} + \bar{v} \times \bar{B}) + (1-\gamma)\frac{\bar{E} \cdot \bar{v}\,\bar{v}}{v^2} \tag{2-13}$$

$$\bar{B}' = \gamma(\bar{B} - \frac{1}{c^2}\,\bar{v} \times \bar{E}) + (1-\gamma)\frac{\bar{B} \cdot \bar{v}\,\bar{v}}{v^2} \tag{2-14}$$

$$\bar{D}' = \gamma(\bar{D} + \frac{1}{c^2}\,\bar{v} \times \bar{H}) + (1-\gamma)\frac{\bar{D} \cdot \bar{v}\,\bar{v}}{v^2} \tag{2-15}$$

$$\bar{H}' = \gamma(\bar{H} - \bar{v} \times \bar{D}) + (1-\gamma)\frac{\bar{H} \cdot \bar{v}\,\bar{v}}{v^2} \tag{2-16}$$

The group of Maxwell equations (2-11)-(2-12), then transforms covariantly to:

$$\nabla' \times \bar{E}' = -\frac{\partial \bar{B}'}{\partial t'} \;;\; \nabla' \cdot \bar{B}' = 0 \tag{2-17}$$

$$\nabla' \times \bar{H}' = \bar{J}' + \frac{\partial \bar{D}'}{\partial t'} \;;\; \nabla' \cdot \bar{D}' = \rho' \tag{2-18}$$

The inverse Lorentz transformations of (2-7)-(2-10) and (2-13)-(2-16) may be found by interchanging the S and the S' coordinates and replacing $\bar{v}$ by $(-\bar{v})$.

2.3 Wave Four-Vector Transformations

For a homogeneous medium, the plane wave solutions in the rest system S of the Maxwell equations (2-11)-(2-12) may be given by:

$$\bar{E}(\bar{r},t) = \text{Re}\,\bar{E}_o e^{-i(\bar{k} \cdot \bar{r} - \omega t)} \tag{2-19}$$

$$\bar{B}(\bar{r},t) = \text{Re}\,\bar{B}_o e^{-i(\bar{k} \cdot \bar{r} - \omega t)} \tag{2-20}$$

where $\bar{E}_o$ and $\bar{B}_o$ are related constants and Re represents "the real part of." The exponential variation contains the wave vector $\bar{k}$ and the angular frequency ω.

Similarly, in the moving system S', the plane wave solution of the Maxwell equations (2-17)-(2-18) may be given by:

$$\bar{E}'(\bar{r}'.t') = \mathrm{Re}\ \bar{E}'_o \ell^{-i(\bar{k}' \cdot \bar{r}' - \omega' t')} \tag{2-21}$$

$$\bar{B}'(\bar{r}',t') = \mathrm{Re}\ \bar{B}'_o \ell^{-i(\bar{k}' \cdot \bar{r}' - \omega' t')} \tag{2-22}$$

where the transformations of space-time are given by (2-7)-(2-8), and the transformations of the fields by (2-13)-(2-14).

Substituting (2-19) and (2-20) in (2-13) and comparing the resulting expression with (2-21), one finds [Papas, 1965]:

$$\bar{E}'_o = \gamma(\bar{E}_o + \bar{v} \times \bar{B}_o) + (1-\gamma)\frac{\bar{E}_o \cdot \bar{v}\,\bar{v}}{v^2} \tag{2-23}$$

$$\bar{k}' \cdot \bar{r}' - \omega' t' = \bar{k} \cdot \bar{r} - \omega t = \emptyset \tag{2-24}$$

where $\emptyset$ is the phase of the wave, and it is an invariant quantity under the Lorentz transformation. In order to find $\bar{k}'$, ω' in terms of $\bar{k}$, ω, it is noticed that the invariant quantity in (2-24) is a scalar product of the two 4-vectors ($\bar{r}$, ict) and ($\bar{k}$, i ω/c). The latter is called the wave 4-vector, and its first three components k_x, k_y, k_z represent the space wave-vector while the fourth component contains the angular frequency ω of the wave. A 4-vector must transform as (2-7)-(2-8), and hence one can immediately write:

$$\bar{k}' = \bar{k} - \frac{\gamma\omega}{c^2}\bar{v} + (\gamma-1)\frac{\bar{k} \cdot \bar{v}}{v^2}\bar{v} \tag{2-25}$$

$$\omega' = \gamma(\omega - \bar{v} \cdot \bar{k}) \tag{2-26}$$

and the inverse transformation:

$$\bar{k} = \bar{k}' + \frac{\gamma\omega'\bar{v}}{c^2} + (\gamma-1)\frac{\bar{k}' \cdot \bar{v}\,\bar{v}}{v^2} \tag{2-27}$$

$$\omega = \gamma(\omega' + \bar{v} \cdot \bar{k}') \tag{2-28}$$

The principle of phase invariance (2-24) applies to homogeneous media only, including anisotropic and dispersive homogeneous media [Papas, 1965].

2.4 Plasma-Parameters Transformations

In the present study of moving plasmas we shall often need to transform the plasma parameters observed in the moving system S' to the ones observed in the laboratory system S, in order to obtain results in the system S. Plasma-parameters transformations have been considered by Getmantsev and Rapoport [1960], Scarf [1961], and Unz [1966a]. Scarf's results, however, do not agree with the results of the other two papers, and the application of his results to the refractive index equations for moving plasmas leads to incorrect equations, which are in disagreement with other works on moving plasmas [Chawla and Unz, 1967b]. His statement that the plasma frequency, the cyclotron frequency, the collision frequency, and resonant frequency all undergo Doppler shift under transformation from S' to S is not correct, since all these frequencies are the natural frequencies of the plasma. Unlike the electromagnetic wave frequencies, the natural frequencies of the plasma do not have a wave vector associated with them, and thus, do not have the property of phase invariance, which results in a Doppler shift. In the present section the relativistic transformations of the various related plasma parameters and natural frequencies will be derived.

Let us assume that the rest system of coordinates S' is moving with a moving plasma at velocity $\underline{\bar{v}}$ with respect to the laboratory system of coordinates S. The corresponding plasma frequencies ω_p', ω_p of the same moving plasma in the respective frames will be defined [Budden, 1961] in the form:

$$\omega_p'^2 = \frac{e'^2 N'}{\epsilon_o m'} \quad ; \quad \omega_p^2 = \frac{e^2 N}{\epsilon_o m} \tag{2-29}$$

where (-e', m') and (-e, m) are the electron charge and mass in the respective frames, N', N are the electron number (volume) density as seen in the respective frames, and ϵ_o is the free space permittivity.

From the special theory of relativity it is well known that the mass of a moving particle increases while its charge remains invariant:

$$e' = e;\ m' = m/\gamma \tag{2-30}$$

where $\gamma = (1-\beta^2)^{-1/2}$ and $\beta = v/c$. As a result of the relativistic Lorentz-Fitzgerald contraction of a moving rod in the direction of motion, one has the following transformation for the volume of a parallelepiped [Pauli, 1958]:

$$V' = \gamma V \tag{2-31}$$

where V', V are the volumes as seen in the respective frames. Since the total number of electrons in V' as counted in the S' moving frame of reference is the same as the total number of electrons in V as counted in the S laboratory frame of reference, one has for N', N, the corresponding electron number volume densities (number/volume), the following relationship by using (2-31):

$$N' = N/\gamma \tag{2-32}$$

Substituting (2-30) and (2-32) in (2-29), one obtains [Unz, 1966a]:

$$\omega_p' = \omega_p \tag{2-33}$$

The plasma frequency is invariant under the Lorentz transformation.

Taking the average counted number of collisions of an electron with heavier particles in the rest frame of reference S' during an interval of time $\Delta t'$, which corresponds to the same average counted number of collisions in the laboratory frame of reference S during a corresponding interval of time Δt, the corresponding collision frequencies ν', ν will be defined as follows:

$$\nu' = \frac{\text{No. Collisions}}{\Delta t'}\ ;\ \nu = \frac{\text{No. Collisions}}{\Delta t} \tag{2-34}$$

In accordance with the Einstein relativistic time dilatation effect, the shortest time interval is shown by the proper clock in the proper system [Pauli, 1958] as follows:

$$\Delta t' = \Delta t/\gamma \tag{2-35}$$

Substituting (2-35) in (2-34), one obtains [Unz, 1966a]:

$$\nu' = \gamma\nu \tag{2-36}$$

The corresponding gyromagnetic (cyclotron) frequencies ω_H', ω_H in the frame of references S' and S will be defined [Budden, 1961] as follows:

$$\omega_H' = \frac{e'B_o'}{m'} \; ; \; \omega_H = \frac{eB_o}{m} \tag{2-37}$$

Assuming that we have a neurtal moving plasma with no net electrostatic field in the laboratory coordinate system S, we take $\overline{E}_o = 0$ in (2-14) and obtain:

$$\overline{B}_o' = \gamma\overline{B}_o + (1-\gamma)\frac{\overline{B}_o\cdot\overline{v}\;\overline{v}}{v^2} \tag{2-38}$$

Substituting (2-30) and (2-38) in (2-37), one obtains [Unz, 1968]:

$$\overline{\omega}_H' = \gamma^2\overline{\omega}_H + \gamma(1-\gamma)\frac{\overline{\omega}_H\cdot\overline{v}}{v^2}\overline{v} \tag{2-39}$$

Taking the static magnetic field in (2-39) along the velocity direction, one obtains [Unz, 1966a] for the longitudinal case:

$$\omega_{HL}' = \gamma\omega_{HL} \tag{2-40}$$

Taking the static magnetic field in (2-39) normal to the velocity direction, one obtains for the transverse case:

$$\omega_{HT}' = \gamma^2\omega_{HT} \tag{2-41}$$

The corresponding acoustic velocity in the electron gas a', a in the frames of reference S' and S will be defined [Unz, 1966b] as follows.

$$a'^2 = \frac{\alpha p'}{m'N'} \; ; \; a^2 = \frac{\alpha p}{mN} \tag{2-42}$$

where p', p are the corresponding scalar pressures, N', N are the corresponding electron number (volume) densities, and α is the ratio of the electron gas specific heat at constant pressure to its specific heat at constant volume and is considered to be a universal constant.

Under the Lorentz transformation the scalar pressure is invariant [Pauli, 1958]:

$$p' = p \tag{2-43}$$

From entropy considerations, the Lorentz transformation of temperature is given in the form [Pauli, 1958];

$$T' = \gamma T \tag{2-44}$$

The equation of state of the perfect electron gas in the rest frame of reference S' is given by:

$$p' = KN'T' \tag{2-45}$$

where K is the universal Boltzmann constant. Substituting (2-32), (2-43), and (2-44) in (2-45), one obtains for the laboratory frame of reference S:

$$p = KNT \tag{2-46}$$

Comparing (2-45) and (2-46), one finds that "The equation of state of an ideal gas remains unchanged in relativistic mechanics" [Pauli, 1958], in accordance with Einstein's first postulate of the special theory of relativity. Substituting (2-30), (2-32), and (2-43) in (2-42), one obtains [Chawla and Unz, 1966b]:

$$a' = \gamma a \tag{2-47}$$

The plasma-parameters relativistic transformations between the rest system S' embedded in the moving plasma and the laboratory system of coordinates S have been given in (2-33), (2-36), (2-39) and (2-47) and will be used later in the book.

2.5 Two Theories of Moving Media

In this section we shall summarize the well-known Minkowski theory of moving media [Sommerfeld, 1964a] and the recently developed Chu theory of moving media [Fano, Chu, Adler, 1960]. The latter has been shown by Tai [1964a, 1967] to be derivable from the former.

The basis of the Minkowski theory of polarizable and magnetizable moving media is the covariance of the Maxwell equations in the form (2-11)-(2-12) and (2-17)-(2-18), which contain the four field vectors $\overline{E}$, $\overline{H}$, $\overline{D}$, $\overline{B}$. In the moving frame S', one has

$$\overline{D}' = \epsilon'\overline{E}' \tag{2-48}$$

$$\overline{B}' = \mu'\overline{H}' \tag{2-49}$$

After using the field transformations (2-13)-(2-16), these constitutive relations in the laboratory system then become

$$\overline{D} + \frac{1}{c^2}\,\overline{v} \times \overline{H} = \epsilon'(\overline{E} + \overline{v} \times \overline{B}) \tag{2-50}$$

$$\overline{B} - \frac{1}{c^2}\,\overline{v} \times \overline{E} = \mu'(\overline{H} - \overline{v} \times \overline{D}) \tag{2-51}$$

which are the constitutive relations of the Minkowski theory. The vectors $\overline{B}$ and $\overline{D}$ in (2-50)-(2-51) can be written in terms of $\overline{E}$ and $\overline{H}$ only [Tai, 1967]:

$$\overline{D} = \epsilon'\overline{\overline{\alpha}}\cdot\overline{E} + \overline{\Omega} \times \overline{H} \tag{2-52}$$

$$\overline{B} = \mu'\overline{\overline{\alpha}}\cdot\overline{H} - \overline{\Omega} \times \overline{E} \tag{2-53}$$

where $\overline{\Omega} = \dfrac{(n^2-1)}{(1-n^2\beta^2)c}\,\hat{z}$; $n = (\mu'\epsilon'/\mu_o\epsilon_o)^{\frac{1}{2}}$;

$$\overline{\overline{\alpha}} = \begin{bmatrix} a & 0 & 0 \\ 0 & a & 0 \\ 0 & 0 & 1 \end{bmatrix} ; \quad a = \frac{1-\beta^2}{1-n^2\beta^2} ; \quad \beta = v/c.$$

Defining the polarization and magnetization vectors $\overline{P}$ and $\overline{M}$ so that in the coordinate system S:

$$\bar{D} = \epsilon_0 \bar{E} + \bar{P} \qquad (2\text{-}54)$$

$$\bar{B} = \mu_0(\bar{H} + \bar{M}) \qquad (2\text{-}55)$$

and in the coordinate system S':

$$\bar{D}' = \epsilon_0 \bar{E}' + \bar{P}' \qquad (2\text{-}56)$$

$$\bar{B}' = \mu_0(\bar{H}' + \bar{M}'), \qquad (2\text{-}57)$$

one can find the following relations:

$$\bar{M} = \bar{\bar{\gamma}} \cdot (\bar{M}' - \bar{v} \times \bar{P}') \qquad (2\text{-}58)$$

$$\bar{P} = \bar{\bar{\gamma}} \cdot (\bar{P}' + \frac{1}{c^2} \bar{v} \times \bar{M}') \qquad (2\text{-}59)$$

where

$$\bar{\bar{\gamma}} = \begin{bmatrix} \gamma & 0 & 0 \\ 0 & \gamma & 0 \\ 0 & 0 & 1 \end{bmatrix}; \quad \gamma = (1-\beta^2)^{-\frac{1}{2}}; \quad \beta = \frac{v}{c}$$

The equations (2-58) and (2-59) show that for a medium which is polarizable only in the system S'($\bar{M}'=0$), there is magnetization observed in the laboratory system ($\bar{M} \neq 0$). This result is rather striking and is not consistent with the Chu formulation.

In the Chu formulation, one only has the two field vectors $\bar{E}$ and $\bar{H}$. The $\bar{E}_c$ and $\bar{H}_c$ in the Chu formulation are related to the $\bar{E}_M$ and $\bar{H}_M$ in the Minkowski formulation by:

$$\bar{E}_M = \bar{E}_c - \mu_0 \bar{v} \times \bar{M}_c; \quad \bar{H}_M = \bar{H}_c + \bar{v} \times \bar{P}_c \qquad (2\text{-}60)$$

where

$$\bar{P}_c = \bar{\bar{\gamma}} \cdot \bar{P}' = (\epsilon' - \epsilon_0) \bar{\bar{\gamma}} \cdot \bar{\bar{\gamma}} \cdot (\bar{E}_c + \mu_0 \bar{v} \times \bar{H}_c) \qquad (2\text{-}61)$$

$$\bar{M}_c = \bar{\bar{\gamma}} \cdot \bar{M}' = (\mu' - \mu_0) \bar{\bar{\gamma}} \cdot \bar{\bar{\gamma}} \cdot (\bar{H}_c - \epsilon_0 \bar{v} \times \bar{E}_c) \qquad (2\text{-}62)$$

The equations (2-61)-(2-62) show that the magnetization $\bar{M}_c$ will be identically zero in the system S if the medium is only polarizable in the moving system S', and as mentioned above, it is in disagreement with the Minkowski theory. However, the difference can be understood in terms of the difference between the definitions of $\bar{H}_c$ and $\bar{H}_M$ [Tai, 1964a]. The Maxwell equations in the Chu formulation are written as:

$$\nabla \times \bar{E}_c + \mu_0 \frac{\partial \bar{H}_c}{\partial t} = -\mu_0 \frac{\partial \bar{M}_c}{\partial t} + \mu_0 \nabla \times (\bar{v} \times \bar{M}_c) \tag{2-63}$$

$$\nabla \times \bar{H}_c - \epsilon_0 \frac{\partial \bar{E}_c}{\partial t} = \frac{\partial \bar{P}_c}{\partial t} - \nabla \times (\bar{v} \times \bar{P}_c) + \bar{J} \tag{2-64}$$

For non-magnetizable medium, one takes $\bar{M}_c = 0$.

III

THE MAGNETO-IONIC THEORY FOR MOVING TEMPERATE PLASMA

3.1 Introduction

The magneto-ionic theory is the theory which deals with the propagation of electromagnetic waves passing through a temperate gas of neutral molecules, in which is imbedded a statistically homogeneous mixture of free electrons and neutralizing heavy positive ions in the presence of an imposed uniform magneto-static field. The magneto-ionic theory and its applications to the stationary ionosphere have been given in two excellent books by Ratcliffe [1959] and Budden [1961]. The complex refractive index and the wave polarizations for electromagnetic waves propagating in the $\hat{z}$ direction in a homogeneous, temperate, stationary non-moving plasma with an imposed uniform magneto-static field are given by the Appleton-Hartree equations [Ratcliffe, 1959; Budden, 1961].

The movements of the ionosphere were considered by Martyn, Maeda, Ratcliffe, and several others. A list of references was given by Unz [1961]. The propagation of electromagnetic waves in a magneto-plasma moving with a constant velocity was first discussed by Bailey [1948, 1950, 1951]. Bell and Helliwell [1960] discussed the amplifications of the whistler mode of propagation in the ionosphere. The magneto-ionic theory for moving temperate plasma was developed independently by Unz [1962], and his theory was corrected by Epstein and Unz [1963] and by Bell, Smith, Brice, and Unz [1963]. This work was extended later to the case of oblique incidence [Unz, 1965a]. An extensive review of the electrodynamics of moving media was given by Tai [1964a], who extended Minkowski's theory of moving media to the anisotropic case [Tai, 1965a] and applied it to the magneto-ionic theory for moving plasma.

The model used by Bailey [1948] is known as the convection current model (CCM), where free space is pervaded by plasma moving uniformly. The model used by Unz [1962, 1965a] was based on Chu's formulation for moving media [Fano, Chu, and Adler, 1960] and is known as the polarization current model (PCM). It may be shown that the refractive index equation derived by Bailey [1948] using the convective current model (CCM) is identical with the result derived by Unz [1965a] using the polarization current model (PCM). The equivalence between the CCM-model and PCM-model was discussed by Tai [1965a], who derived the same refractive index equation by using his Maxwell-Minkowski equations for small velocities for the anisotropic case.

In the present chapter we shall first present the general equations and discuss the longitudinal case, sometimes called the whistler mode of propagation. The magneto-ionic theory for moving temperate plasma will be derived for the non-relativistic general case by using the polarization current model (PCM). An algebraic equation for the complex refractive index will be found and it will be discussed for several special cases.

Throughout the analysis it is assumed that the constant drift velocity $\overline{v}_0$ of the plasma and the static magnetic field $\overline{H}_0$ are completely independent of each other. Physically this is the case if they are both in the same direction, as in the longitudinal propagation case. In case they are not parallel to each other, a transverse constant electric field could be applied to make them independent, and this static electric field will not affect the basic perturbation relationships which will be used here.

In the present chapter the harmonic time-varing wave propagation in the longitudinal $\hat{z}$ direction will be described by the factor $e^{i(\omega t - k_0 nz)}$, where $k_0 = \omega/c$ and $n = c/u$ is the refractive index, u being the phase velocity of the electromagnetic wave in the plasma. In general, the refractive index will be complex $n = \mu - i\chi$, μ being the wave refractive index and χ representing the absorption of the wave [Budden, 1961]. The charge of the electron will be taken as (-e), where e will have a positive numerical value.

3.2 The General Basic Equations

The macroscopic Maxwell equations in a moving matter with a drift velocity $\overline{v}_0$ could be written [Fano, Chu, and Adler, 1960] as follows:

$$\nabla \times \overline{H} - \epsilon_0 \frac{\partial \overline{E}}{\partial t} = \overline{J}_f + \frac{\partial \overline{P}}{\partial t} + \nabla \times (\overline{P} \times \overline{v}_0) \tag{3-1}$$

$$\nabla \times \overline{E} + \mu_0 \frac{\partial \overline{H}}{\partial t} = -\frac{\partial}{\partial t}(\mu_0 \overline{M}) - \nabla \times (\mu_0 \overline{M} \times \overline{v}_0) \tag{3-2}$$

$$\nabla \cdot (\epsilon_0 \overline{E}) = \rho_f - \nabla \cdot \overline{P} \tag{3-3}$$

$$\nabla \cdot (\mu_0 \overline{H}) = -\nabla \cdot (\mu_0 \overline{M}) \tag{3-4}$$

where ϵ_0, μ_0 are the permittivity and permeability of free space. Assuming the continuity equations and taking only dynamic fields, equations (3-1) and (3-2) will be sufficient for our present consideration. They are similar to (2-63) and (2-64) with the subscript c omitted here.

Taking the polarization current model (PCM) for the plasma [Unz, 1962, 1965a], one has no free electric current [Budden, 1961] $\overline{J}_f = 0$, and no magnetic polarization M = 0, and (3-1) and (3-2) become:

$$\nabla \times \overline{H} - \epsilon_0 \frac{\partial \overline{E}}{\partial t} = \frac{\partial \overline{P}}{\partial t} + \nabla \times (\overline{P} \times \overline{v}_0) \tag{3-5}$$

$$\nabla \times \overline{E} + \mu_0 \frac{\partial \overline{H}}{\partial t} = 0 \tag{3-6}$$

If there are N electrons per unit volume and all are displaced by a distance $\overline{R}$, the equivalent dipole moment per unit volume, or the polarization $\overline{P}$, is given by [Ratcliffe, 1959]:

$$\overline{P} = -eN\overline{R} \tag{3-7}$$

The equation of motion of an electron of mass m and charge (-e) is given in general by:

$$m \frac{d^2\overline{R}}{dt^2} + m\nu \frac{d\overline{R}}{dt} = -e\overline{E}_T - e\mu_0(\overline{v}_T \times \overline{H}_T) \tag{3-8}$$

where ν is the collision frequency and $\bar{E}_T$, $\bar{H}_T$, $\bar{v}_T$ are the total electric field, total magnetic field, and total velocity of the electron, respectively. Each one consists of a constant field component $\bar{E}_o$, $\bar{H}_o$, $\bar{v}_o$ and a harmonic time-varying component of the wave $\bar{E}$, $\bar{H}$, $\bar{P}$ as follows:

$$\bar{E}_T = \bar{E}_o + \bar{E}e^{+i(\omega t - k_o nz)} \tag{3-9}$$

$$\bar{H}_T = \bar{H}_o + \bar{H}e^{+i(\omega t - k_o nz)} \tag{3-10}$$

$$\bar{v}_T = \bar{v}_o + \frac{d}{dt}\left[\bar{R}_1 e^{+i(\omega t - k_o nz)}\right] \tag{3-11}$$

In (3-8) and (3-11) one should use the total derivative or the material derivative:

$$\frac{d\bar{R}}{dt} = \frac{D\bar{R}}{Dt} = \left(\frac{\partial}{\partial t} + \bar{v}_o\cdot\nabla\right)\bar{R} \tag{3-12}$$

Taking $\bar{R} = \bar{R}_1 e^{i(\omega t - k_o nz)}$ in (3-12) where R_1 = constant, one has for the present case:

$$\frac{d\bar{R}}{dt} = \frac{D\bar{R}}{Dt} = i\omega(1-n\beta_z)\bar{R} = i\omega s\bar{R} \tag{3-13}$$

where $s = 1-n\beta_z$. It is assumed that the harmonic time-varying components of the wave in (3-9)-(3-11) are much smaller than the corresponding constant components. This is being done in accordance with the perturbational approach of the small-signal theory in electron tubes. The double frequency components will be neglected since they are of the second order of magnitude. Substituting (3-9) - (3-13) in (3-8) and using small-signal theory, equation (3-8) could be linearized to give:

$$-m\omega^2 s^2\bar{R} + m\nu i\omega s\bar{R} = -e\bar{E} - e\mu_o i\omega s(\bar{R}\times\bar{H}_o) - e\mu_o(\bar{v}_o\times\bar{H}) \tag{3-14}$$

where $\bar{E}$ and $\bar{H}$ are the electromagnetic fields of the wave, $\bar{H}_o$ is the imposed static magnetic field, and $\bar{v}_o$ is the constant drift velocity of the plasma.

Let us define [Budden, 1961] in general:

$$X = \frac{\omega_p^2}{\omega^2} = \frac{Ne^2}{\epsilon_o m\omega^2} \tag{3-15}$$

$$\bar{Y} = \frac{\bar{\omega}_H}{\omega} = \frac{\mu_o e\bar{H}_o}{m\omega} = Y_x\hat{x} + Y_y\hat{y} + Y_z\hat{z} \tag{3-16}$$

$$Z = \frac{\nu}{\omega} \tag{3-17}$$

$$\bar{\beta} = \frac{\bar{v}_o}{c} = \beta_x\hat{x} + \beta_y\hat{y} + \beta_z\hat{z} \tag{3-18}$$

It should be pointed out that in our present notation the charge of the electron is (-e) where e is a positive number. In our definition in (3-16), we have actually $\bar{Y} = +(\mu_o|e|/m\omega)\bar{H}_o$ where $\bar{Y}$ and $\bar{H}_o$ are in the same direction. Budden [1961] defines $\bar{Y} \triangleq -(\mu_o|e|/m\omega)\bar{H}_o$, where $\bar{Y}$ and $\bar{H}_o$ are in opposite directions. Therefore the sign of $\bar{Y}$ in the present book will be opposite when compared with the sign of $\bar{Y}$ given by Budden [1961] for the corresponding formulas.

Substituting (3-7) in (3-14) and using definitions (3-15) - (3-18), one obtains in general:

$$\epsilon_o X[\bar{E} + (\bar{\beta} \times \eta_o\bar{H})] = -s\, U\bar{P} - is(\bar{Y} \times \bar{P}) \tag{3-19}$$

where $\eta_o = \sqrt{\mu_o/\epsilon_o}$ and $U = s - i\,Z$. Equation (3-19) represents the general constitutive relation for the drifting plasma.

The Maxwell equations (3-5) and (3-6) and the constitutive relation (3-19) represent three vector equations with three vector unknowns, $\bar{E}$, $\bar{H}$, and $\bar{P}$, and will be used as the general basic equations for the discussion of the propagation of the electromagnetic waves in drifting magneto-plasma in the next two sections.

3.3 The Longitudinal Propagation Case

In the longitudinal propagation case, which is sometimes called the whistler mode of propagation, the constant drift velocity of the plasma, the imposed static magnetic field, and the direction of the electromagnetic wave propagation are all in the same longitudinal direction. Taking the plasma constant drift velocity to be $\bar{v}_o = v_o\hat{z}$, where $\hat{z}$ is a unit vector, and assuming that the plane wave propagation in the longitudinal $\hat{z}$ direction will be described by the common factor $e^{i(\omega t - k_o nz)}$ in all the wave components, one will obtain from (3-5) the following relations in terms of rectangular components:

$$ik_0 n\, H_y - i\omega\epsilon_0 E_x = i\omega P_x - ik_0 n\, v_0\, P_x \tag{3-20}$$

$$-ik_0\, n\, H_x - i\omega\epsilon_0 E_y = i\omega P_y - ik_0\, n\, v_0 P_y \tag{3-21}$$

$$D_z = \epsilon_0 E_z + P_z = 0 \tag{3-22}$$

Taking $c = 1/\sqrt{\mu_0\epsilon_0}$ and $\eta_0 = \sqrt{\mu_0/\epsilon_0}$, one similarly obtains from (3-6):

$$\eta_0 H_x = -n\, E_y \tag{3-23}$$

$$\eta_0 H_y = +n\, E_x \tag{3-24}$$

$$H_z = 0 \tag{3-25}$$

From (3-25) one sees that the magnetic field vector of the plane wave lies entirely in the wave front and thus represents a TM wave. The electric field vector may have a component perpendicular to the wave front in the direction of propagation.

Substituting (3-23) and (3-24) in (3-20) and (3-21), taking $\beta_L = v_0/c$ and $s = 1 - n\beta_L$, one obtains:

$$\epsilon_0(n^2 - 1)E_x = sP_x \tag{3-26}$$

$$\epsilon_0(n^2 - 1)E_y = sP_y \tag{3-27}$$

From (3-23) - (3-27) one may define the polarization of the wave ρ in the form:

$$\rho = \frac{E_y}{E_x} = \frac{P_y}{P_x} = -\frac{H_x}{H_y} \tag{3-28}$$

Taking for the present longitudinal propagation case $\overline{Y} = Y_L\hat{z}$, $\overline{\beta} = \beta_L\hat{z}$ and substituting (3-23) - (3-25) in the constitutive relation (3-19), after rewriting it in terms of rectangular components, one obtains:

$$\epsilon_0 X\, E_x = -UP_x + i\, Y_L\, P_y \tag{3-29}$$

$$\epsilon_0 X\, E_y = -i\, Y_L\, P_x - UP_y \tag{3-30}$$

$$\epsilon_o X E_z = -s U P_z \tag{3-31}$$

where (3-29) - (3-31) represent the constitutive relations for the present longitudinal propagation case.

Substituting (3-22) in (3-31), one has:

$$s U - X = 0 \tag{3-32}$$

Taking $U = s - iZ$ and $s = 1 - n\beta_L$ in (3-32), one has:

$$(1 - n\beta_L)^2 - i Z(1 - n\beta_L) - X = 0 \tag{3-33}$$

where (3-33) represents the independent space charge modes and degenerates to the resonant plasma frequency $X = 1$ or $\omega = \omega_p$ for no plasma drift ($\beta_L = 0$) and no collisions ($Z = 0$).

Substituting (3-29) and (3-30) in (3-26) and (3-27), one obtains:

$$[U(n^2 - 1) + s X] P_x - i Y_L (n^2 - 1) P_y = 0 \tag{3-34}$$

$$+ i Y_L (n^2 - 1) P_x + [U(n^2 - 1) + sX] P_y = 0 \tag{3-35}$$

For a non-trivial solution, the determinant of the coefficients of (3-34) and (3-35) should be equated to zero, and thus one obtains:

$$U(n^2 - 1) + sX = \mp Y_L (n^2 - 1) \tag{3-36}$$

which may be rewritten in the form:

$$(n^2 - 1)[U \pm Y_L] + sX = 0 \tag{3-37}$$

Taking $U = s - iZ$ and $s = 1 - n\beta_L$, (3-37) may be written explicitly in the form:

$$(n^2 - 1)[1 - n\beta_L - iZ \pm Y_L] + (1 - n\beta_L)X = 0 \tag{3-38}$$

The cubic equation (3-38) represents the refractive index equation in an infinite plasma for the longitudinal whistler mode of propagation, where $+ Y_L$ refers to the ordinary wave and $- Y_L$ refers to the extraordinary wave. Some numerical results of (3-38) have been given by Sidhu and Unz [1968].

Using definition (3-28) for the wave polarization ρ, one obtains from (3-34) and (3-35):

$$\rho = -\frac{1}{\rho} \; ; \; \rho = \mp i \tag{3-39}$$

Equation (3-39) shows that the electromagnetic wave for the longitudinal propagation case is circularly polarized, and this result is identical with the classical magneto-ionic theory for stationary plasma [Ratcliffe, 1959; Budden, 1961].

For the particular case of no collisions $Z = 0$ and no magneto-static field $Y_L = 0$, (3-38) becomes:

$$n^2 = 1 - X \tag{3-40}$$

The propagation of electromagnetic waves in drifting isotropic plasma with no collisions is independent of the plasma drift velocity.

The total reflection of the electromagnetic wave in the drifting plasma will occur at the level $n = 0$, and for no collision $Z = 0$, (3-38) becomes:

$$X = 1 \pm Y_L \tag{3-41}$$

Equation (3-41) is identical with the result for stationary plasma [Ratcliffe, 1959; Budden, 1961].

The same identical refractive index equation (3-38) for the longitudinal propagation case applies also for the relativistic case [Unz, 1966a] for a large plasma drift velocity. The details of the derivation for the relativistic case are given in section 6.3.

3.4 The Non-Relativistic Solution

The aim of the present section is to develop the magneto-ionic theory for moving temperate plasma for the general case, where the plasma drift velocity $\overline{v}_o$ and the imposed static magnetic field $\overline{H}_o$ are in arbitrary directions with respect to the electromagnetic wave propagating in the $\hat{z}$ direction. Assuming that the plane wave propagation in the $\hat{z}$ direction will be described by the common factor $e^{i(\omega t - k_o nz)}$ in all the wave components and that the plasma drift velocity is given by $\overline{v}_o = v_{ox}\hat{x} + v_{oy}\hat{y} + v_{oz}\hat{z}$, where $\hat{x}$, $\hat{y}$, $\hat{z}$ are the corresponding unit vectors, one obtains from (3-5) the following relations in terms of rectangular components:

$$i k_o n H_y - i\omega\epsilon_o E_x = i\omega P_x - i k_o n v_{oz} P_x + i k_o n v_{ox} P_z \qquad (3\text{-}42)$$

$$-i k_o n H_x - i\omega\epsilon_o E_y = i\omega P_y - i k_o n v_{oz} P_y + i k_o n v_{oy} P_z \qquad (3\text{-}43)$$

$$D_z = \epsilon_o E_z + P_z = 0 \qquad (3\text{-}44)$$

The corresponding result from (3-6) is given in (3-23)-(3-25). Substituting (3-23) and (3-24) in (3-42) and (3-43) and defining $\beta_x = v_{ox}/c$, $\beta_y = v_{oy}/c$, $\beta_z = v_{oz}/c$, and $s = 1 - n\beta_z$, one obtains:

$$\epsilon_o(n^2 - 1)E_x = sP_x + n\beta_x P_z \qquad (3\text{-}45)$$

$$\epsilon_o(n^2 - 1)E_y = sP_y + n\beta_y P_z \qquad (3\text{-}46)$$

$$\epsilon_o E_z + P_z = 0 \qquad (3\text{-}47)$$

Writing the general constitutive relation (3-19) in terms of rectangular components and substituting (3-23)-(3-24), one obtains after rearranging:

$$\epsilon_o X E_x = -UP_x + i Y_z P_y - i Y_y P_z \qquad (3\text{-}48)$$

$$\epsilon_o X E_y = -i Y_z P_x - UP_y + i Y_x P_z \qquad (3\text{-}49)$$

$$\epsilon_o X(E_z + n\beta_x E_x + n\beta_y E_y) = i s Y_y P_x - i s Y_x P_y - s UP_z \qquad (3\text{-}50)$$

Equations (3-45)-(3-50) give us six homogeneous equations with six unknowns. For a non-trivial solution, the determinant of the coefficients will have to be equated to zero, and this will give the algebraic equation for the refractive index equation for the characteristic waves. In order to simplify the determinant from a 6 x 6 determinant to a 3 x 3 determinant, the electric field components may be eliminated to obtain three homogeneous equations with three unknowns.

Multiplying both sides of (3-48) - (3-50) by $(n^2 - 1)$ and substituting (3-45) - (3-47), one obtains:

$$A P_x - i D_z P_y + (B_x + i D_y) P_z = 0 \qquad (3\text{-}51)$$

$$i D_z P_x + A P_y + (B_y - i D_x) P_z = 0 \qquad (3\text{-}52)$$

$$s(B_x - i\,D_y)\,P_x + s\,(B_y + i\,D_x)\,P_y + C\,P_z = 0 \qquad (3\text{-}53)$$

where we define:

$$A = U(n^2 - 1) + Xs \qquad (3\text{-}54)$$

$$B_x = X\,\beta_x\,n\ ;\ B_y = X\,\beta_y\,n \qquad (3\text{-}55)$$

$$C = (n^2 - 1)\,(s\,U - X) + X\,n^2\,(\beta_x^2 + \beta_y^2) \qquad (3\text{-}56)$$

$$D_x = (n^2 - 1)Y_x;\ D_y = (n^2 - 1)Y_y;\ D_z = (n^2 - 1)Y_z \qquad (3\text{-}57)$$

$$U = s - i\,Z\ ;\ s = 1 - n\,\beta_z \qquad (3\text{-}58)$$

For a non-trivial solution of (3-51) - (3-53), one has:

$$\begin{vmatrix} A & -iD_z & B_x+i\,D_y \\ +i\,D_z & A & B_y - i\,D_x \\ s(B_x - i\,D_y) & s(B_y + i\,D_x) & C \end{vmatrix} = 0 \qquad (3\text{-}59)$$

The determinantal equation (3-59) gives us the equation for the refractive index n of the characteristic waves in the infinite moving magneto-plasma media.

Developing the determinant in (3-59) and re-arranging, one obtains:

$$C\,(A^2 - D_z^{\,2}) - sA\,(B_x^{\,2} + B_y^{\,2} + D_x^{\,2} + D_y^{\,2}) - 2s\,D_z(B_xD_x + B_yD_y) = 0 \qquad (3\text{-}60)$$

Since the electromagnetic wave propagates in the $\hat{z}$ direction, one can distinguish between longitudinal L and transverse T components as follows:

$$Y_L = Y_z\ ;\ \overline{Y}_T = Y_x\hat{x} + Y_y\hat{y} \qquad (3\text{-}61)$$

$$\beta_L = \beta_z\ ;\ \overline{\beta}_T = \beta_x\hat{x} + \beta_y\hat{y} \qquad (3\text{-}62)$$

From (3-61) and (3-62) one may obtain:

$$Y_T^{\,2} = \overline{Y}_T\cdot\overline{Y}_T = Y_x^{\,2} + Y_y^{\,2} \qquad (3\text{-}63)$$

$$\beta_T^2 = \bar{\beta}_T \cdot \bar{\beta}_T = \beta_x^2 + \beta_y^2 \tag{3-64}$$

$$\bar{\beta}_T \cdot \bar{Y}_T = \beta_x Y_x + \beta_y Y_y \tag{3-65}$$

Substituting (3-54) - (3-58) into (3-60) and using (3-61) - (3-65), one may obtain the final result in the following form:

$$\begin{aligned} &[(n^2 - 1)(s^2 - iZs - X) + X\beta_T^2 n^2]\cdot \\ &\cdot\left\{[(n^2 - 1)(s - iZ) + Xs]^2 - Y_L^2 (n^2 - 1)^2\right\} - \\ &- s[(n^2 - 1)(s - iZ) + Xs]\cdot[Y_T^2 (n^2 - 1)^2 + X^2 \beta_T^2 n^2] - \\ &- 2XY_L n s (n^2 - 1)^2 (\bar{\beta}_T \cdot \bar{Y}_T) = 0 \end{aligned} \tag{3-66}$$

where $s = 1 - n\beta_L$, L and T representing the longitudinal and transverse components, respectively. Equation (3-66) gives the algebraic equation for the refractive index n of the characteristic waves in the magneto-ionic theory for moving temperate plasma. It may be shown that the result found by Bailey [1948], using the convective current model (CCM), is identical with the result (3-66) found here, using the polarization current model (PCM).

3.5 The Non-Relativistic Refractive Index Equation

In the present section, equation (3-66) will be revised and investigated for different particular cases. Collecting the terms involving β_T^2 in (3-66) and rearranging, one obtains:

$$\begin{aligned} &X\beta_T^2 n^2 \left\{[(n^2 - 1)(s - iZ) + Xs]^2 - Y_L^2 (n^2 - 1)^2 - \right. \\ &\left. - Xs[(n^2 - 1)(s - iZ) + Xs]\right\} = \\ &= X\beta_T^2 n^2 (n^2 - 1)\left\{[(n^2 - 1)(s - iZ) + Xs](s - iZ) - \right. \\ &\left. - Y_L^2 (n^2 - 1)\right\} \end{aligned} \tag{3-67}$$

Substituting (3-67) in (3-66) for the β_T^2 terms and cancelling the common factor $(n^2 - 1)$, one obtains:

$$(s^2 - i Z s - X) \left\{ [(n^2 - 1)(s - i Z) + X s]^2 - Y_L^2 (n^2 - 1)^2 \right\} -$$
$$- Y_T^2 s (n^2 - 1) [(n^2 - 1)(s - i Z) + X s] +$$
$$+ X \beta_T^2 n^2 \left\{ [(n^2 - 1)(s - i Z) + X s](s - i Z) - Y_L^2 (n^2 - 1) \right\} -$$
$$- 2 X Y_L n s (n^2 - 1) (\overline{\beta}_T \cdot \overline{Y}_T) = 0 \qquad (3\text{-}68)$$

Equation (3-68) is an algebraic equation of the eighth order in n.

Applying the Minkowski equations to the convective current model, Tai [1965a] obtained a 3 x 3 determinantal equation for the refractive index. This same determinant was derived elsewhere [Epstein and Unz, 1963]. By developing this determinant it is found that the resulting equation for the refractive index is identical with (3-68).

Substituting $s = 1 - n\beta_L$ in the last term of (3-68) and rearranging, one has:

$$[(n^2 - 1)(s - i Z) + X s] \cdot$$
$$\cdot \left\{ [-(s - i Z) + X s](s^2 - i Z s - X) + \right.$$
$$\left. + n^2 s (s - i Z)^2 - X n^2 (s - i Z)(1 - \beta_T^2) \right\} -$$
$$- Y_L^2 (n^2 - 1) \left\{ [s (s - i Z)(n^2 - 1) + X] - \right.$$
$$\left. - X n^2 [1 - \beta_T^2 + 2\beta_L (\overline{\beta}_T \cdot \overline{Y}_T)/Y_L] \right\} -$$
$$- Y_T^2 s (n^2 - 1) [(n^2 - 1)(s - i Z) + X s] -$$
$$- 2 X Y_L n (n^2 - 1)(\overline{\beta}_T \cdot \overline{Y}_T) = 0 \qquad (3\text{-}69)$$

Since the refractive index equations (3-66) - (3-69) have been derived for the non-relativistic case of small drift velocities only, where $\beta_T^2 \ll 1$ and $\beta_L \beta_T \ll 1$ could be neglected (assuming $2\beta_L \overline{\beta}_T \cdot \overline{Y}_T \ll Y_L$ for $Y_L \neq 0$), one could obtain from (3-69) after rearranging:

$$(s^2 - i Z s - X) \cdot$$

$$\cdot \left\{ [(n^2 - 1)(s - i Z) + X s]^2 - Y_L^2 (n^2 - 1)^2 \right\} -$$

$$- Y_T^2 s (n^2 - 1)[(n^2 - 1)(s - i Z) + X s] -$$

$$- 2X Y_L n (n^2 - 1)(\bar{\beta}_T \cdot \bar{Y}_T) = 0 \qquad (3\text{-}70)$$

Equation (3-70) gives the eighth-order algebraic refractive index equation for the non-relativistic magneto-ionic theory for temperate moving plasma.

For the particular case of the whistler mode of propagation $\bar{Y}_T = 0$, one obtains from (3-70):

$$s^2 - i Z s - X = 0 \qquad (3\text{-}71)$$

$$(n^2 - 1)(s - i Z) + X s = \mp Y_L (n^2 - 1) \qquad (3\text{-}72)$$

which are identical with (3-33) and (3-36) for the longitudinal propagation case.

For the particular case of $Y_L = 0$, one obtains form (3-70):

$$(n^2 - 1)(s - i Z) + X s = 0 \qquad (3\text{-}73)$$

$$(s^2 - i Z s - X)[(n^2 - 1)(s - i Z) + X s] - Y_T^2 s (n^2 - 1) = 0 \qquad (3\text{-}74)$$

The total reflection of the electromagnetic wave in drifting magneto-plasma will be at the level $n = 0 (s = 1)$, and for no collisions $Z = 0$ equations (3-68) or (3-70) will thus become:

$$(1 - X)[(1 - X)^2 - (Y_T^2 + Y_L^2)] = 0 \qquad (3\text{-}75)$$

From (3-75) one finds the levels of reflection to be:

$$X = 1; \; X = 1 \pm \sqrt{Y_T^2 + Y_L^2} = 1 \pm Y \qquad (3\text{-}76)$$

Equation (3-76) gives the same levels of reflection for the drifting magneto-plasma as for the stationary non-moving magneto-plasma [Ratcliffe, 1959; Budden, 1961].

For the particular case of stationary magneto-plasma $\beta_T = \beta_L = 0$ and $s = 1$, one obtains from (3-68) or (3-70):

$$(1 - i Z - X)\left\{[(n^2 - 1)(1 - i Z) + X]^2 - Y_L^2 (n^2 - 1)^2\right\} - Y_T^2 (n^2 - 1)[(n^2 - 1)(1 - i Z) + X] = 0 \qquad (3\text{-}77)$$

which could be rewritten in the form:

$$\left[1 - i Z + \frac{X}{n^2 - 1}\right]^2 - \frac{Y_T^2}{1 - i Z - X}\left[1 - i Z + \frac{X}{n^2 - 1}\right] - Y_L^2 = 0 \qquad (3\text{-}78)$$

where the quadratic equation (3-78) could be solved to give the Appleton-Hartree equation [Ratcliffe, 1959; Budden, 1961].

IV

OBLIQUE WAVE PROPAGATION IN MOVING TEMPERATE MAGNETO-PLASMA

4.1 Introduction

The study of wave propagation in any given medium involves the determination of the characteristic equation for the wave vector which the medium can support. In a stationary isotropic dielectric of permittivity ϵ, the magnitude of the wave vector in the direction of propagation is given by $k = (\omega^2 \mu_0 \epsilon)^{\frac{1}{2}}$, where ω is the wave frequency and μ_0 the permeability of free space. For a more complex medium, such as anisotropic dielectric with permittivity given by a tensor $\overline{\overline{\epsilon}}$, for example a stationary plasma in the presence of static magnetic field, one cannot simply put down $k^2 = \omega^2 \mu_0 \cdot \overline{\overline{\epsilon}}$, which is meaningless. To determine the characteristic equation, one usually finds the electric field wave equation from the Maxwell equations. After substituting the factor for the space and the time variation, a system of homogeneous linear equations for the components of the electric field is obtained. The condition for the existence of a non-trivial solution for this system requires that the determinant of the matrix of the coefficients must be zero, and it then leads to the derivation of the characteristic equation.

In the case of magneto-plasmas, the permittivity tensor depends upon the direction and strength of the static magnetic field. Since the wave propagating in the plasma is to be excited by some means, it is assumed that the magneto-plasma is contained in the region $z > 0$ of the configuration space, and that $z = 0$ forms the boundary between the plasma and the un-ionized gas. The boundary is assumed to be sharp in the sense that at the boundary the variation of the free electron density from its value in the plasma to zero takes place in a distance much smaller than one wave length. An electromagnetic wave is incident on the boundary from the region $z < 0$. For normal incidence, where the waves propagate normal to the boundary in the region $z > 0$, the refractive indices are given by the Appleton-Hartree equation [Ratcliffe, 1959; Budden, 1961]. For the case of oblique incidence, the characteristic equation for q is given by the Booker quartic [Booker, 1936, 1939, 1949]. The quantity $q = n \cos\theta$, where n is the refractive index of the wave in the direction of propagation and θ is the angle which the wave makes with the $\hat{z}$ direction, was introduced by Booker [Budden, 1961].

Further complications are introduced when the plasma is assumed to be moving in the region $z > 0$. The physical situation corresponds to a uniformly moving un-ionized gas, which becomes ionized at the plane $z = 0$, or a uniformly moving plasma which is de-ionized at the plane $z = 0$. This problem was first considered by Bailey [1948, 1950] in order to study the excess ncise radiation from the sunspots. Bailey's theory for the amplification was later criticized by Twiss [1951]. The model of the plasma used for the derivation by Bailey, is known today as the convection current model (CCM), where free space is pervaded by neutral gas of charged particles moving uniformly. A different approach was used by Unz [1965a] for the oblique wave propagation in a moving, temperate magneto-plasma, where he found the refractive index equation in a determinantal equation form. The approach used by Unz [1965a] was based on the Chu formulation [Fano, Chu, and Adler, 1960] for moving media, and is presently known as the polarization current model (PCM). The general relativistic refractive index equation for large plasma velocities, for the case of general oblique electromagnetic wave incidence, was derived by Unz [1968].

In this chapter we shall first discuss the plasma polarization current model (PCM) for oblique incident electromagnetic waves. In the next section, Minkowski's theory of moving media will be applied to the same problem, using the convection current model (CCM) of the plasma. It will be shown that the results obtained by the two approaches are identical insofar as the characteristic equation for q is considered. The latter approach was first used by Tai [1965a] for normally incident electromagnetic waves. In the next section the general relativistic refractive index equation for oblique electromagnetic waves in a moving magneto-plasma will be derived [Unz, 1968] and particular cases will be discussed. It will be shown that for the non-relativistic normal incidence case, the different solutions give results which agree with the ones found in chapter three.

4.2 Polarization Current Model (PCM)

With the configuration space described in the previous section, let an incident plane electromagnetic wave have direction cosines S_1, S_2, and C so that the angle θ_i made by the wave direction with the z-axis is given by

$$\cos\theta_i = C;\ \sin\theta_i = (S_1^2 + S_2^2)^{1/2};\ C^2 + S_1^2 + S_2^2 = 1 \tag{4-1}$$

and each field component of the incident wave has the space variation factor:

$$e^{-i\bar{k}\cdot\bar{R}} = e^{-ik_o(S_1x + S_2y + Cz)} \tag{4-2}$$

where the harmonic time variation $e^{+i\omega t}$ is suppressed, $|\bar{k}| = k_o = \omega/c = 2\pi/\lambda$ and λ is the free space wave length. For convenience, it is assumed that the region $z < 0$ is free space and that the plasma is generated at $z = 0$. An extension in order to take into account the dielectric constant of the un-ionized gas, mentioned in the last section, is rather trivial.

The boundary conditions at $z = 0$ require that the x and y dependence in the region $z \geqq 0$ remain the same as in the region $z \leqq 0$. The space variation of the waves in $z \geqq 0$ may therefore be assumed to be of the form:

$$e^{-ik_o\bar{n}\cdot\bar{R}} = e^{-ik_o(S_1x + S_2y + qz)} \tag{4-3}$$

From (4-3) one has:

$$n^2 = S_1^2 + S_2^2 + q^2 \tag{4-4}$$

$$q = n\cos\theta \tag{4-5}$$

where n is the wave refractive index and θ is the angle that the transmitted wave in the plasma makes with the z axis. From (4-1) and (4-4) one has:

$$n^2 = \sin^2\theta_i + q^2 \tag{4-6}$$

From (4-5) and (4-6) one obtains:

$$n \sin\theta = \sin\theta_i \tag{4-7}$$

which is Snell's law for our case. From (4-5) and (4-7) one has:

$$\text{tg } \theta = \frac{\sin\theta_i}{q} \tag{4-8}$$

Once q is calculated, n and θ may be found from (4-6) and (4-8). In the following, the results are obtained in the form of equations in terms of q, which will be the corresponding characteristic equations for the oblique incidence case.

The plasma is assumed to be moving uniformly with arbitrary velocity $\overline{v}_o$, and the static magnetic field $\overline{H}_o$ is assumed to be arbitrarily directed. The two Maxwell equations for polarizable moving media have been given in (3-5) and (3-6) for harmonic time variation $e^{+i\omega t}$:

$$\nabla \times \overline{H}_c - i\omega\epsilon_o\overline{E}_c = i\omega\overline{P} + \nabla \times (\overline{P} \times \overline{v}_o) \tag{4-9}$$

$$\nabla \times \overline{E}_c + i\omega\mu_o\overline{H}_c = 0 \tag{4-10}$$

where $\overline{E}_c$, $\overline{H}_c$ represent the electric and the magnetic fields according to the Chu formulation [Fano, Chu, and Adler, 1960] and $\overline{P}$ is the polarization vector.

Consider a plane electromagnetic wave transmitted through the ionosphere, for which all the wave fields components depend on x, y, z only through the factor given in (4-3). Then we may write symbolically:

$$\frac{\partial}{\partial x} \equiv -i\,k_o\,S_1;\; \frac{\partial}{\partial y} \equiv -\,i\,k_o\,S_2;\; \frac{\partial}{\partial z} \equiv -\,i\,k_o\,q \tag{4-11}$$

Using (4-11), one may rewrite (4-10) in a matrix form as follows:

$$\begin{bmatrix} \eta_o H_{cx} \\ \eta_o H_{cy} \\ \eta_o H_{cz} \end{bmatrix} = \begin{bmatrix} 0 & -q & S_2 \\ q & 0 & -S_1 \\ -S_2 & S_1 & 0 \end{bmatrix} \begin{bmatrix} E_{cx} \\ E_{cy} \\ E_{cz} \end{bmatrix} \tag{4-12}$$

or in the form

$$\eta_o H_c] = [S]\, E_c] \tag{4-13}$$

where $\eta_o = \sqrt{\mu_o/\epsilon_o}$ and [S] is the skew-symmetric matrix in (4-12).

The constant drift velocity $\bar{v}_o$ of the moving magneto-plasma may be defined in terms of its components:

$$\bar{\beta} = \frac{\bar{v}_o}{c} \; ; \; \beta_x = \frac{v_{ox}}{c} ; \; \beta_y = \frac{v_{oy}}{c} \; ; \; \beta_z = \frac{v_{oz}}{c} \tag{4-14}$$

Using (4-11) and (4-14), one may rewrite (4-9) in a matrix form as follows [Unz, 1965a]:

$$[S]\, \eta_o H_c] + [I]\, E_c] = -\frac{1}{\epsilon_o}\left\{ [I] + [S]\, [\beta] \right\} P] \tag{4-15}$$

where [S] is the skew-symmetric matrix in (4-12), [I] is the identity matrix, and the relative plasma velocity skew-symmetric matrix $[\beta]$ is given by:

$$[\beta] = \begin{bmatrix} 0 & -\beta_z & \beta_y \\ \beta_z & 0 & -\beta_x \\ -\beta_y & \beta_x & 0 \end{bmatrix} \tag{4-16}$$

Using definitions (4-12) and (4-16), let us define the matrix $[\Gamma]$ as follows:

$$[\Gamma] = [I] + [\beta]\, [S] = \begin{bmatrix} 1-\beta_y S_2 - \beta_z q & \beta_y S_1 & \beta_z S_1 \\ \beta_x S_2 & 1-\beta_x S_1 - \beta_z q & \beta_z S_2 \\ \beta_x q & \beta_y q & 1-\beta_x S_1 - \beta_y S_2 \end{bmatrix} \tag{4-17}$$

The transposed matrix $[\Gamma]_t$ may be found from matrix $[\Gamma]$ by interchanging rows and columns. Using definitions (4-12) and (4-16), one has [Sokolnikoff, 1951]:

$$[\Gamma]_t = [I]_t + \Big\{ [\beta][S] \Big\}_t = [I] + [S]_t [\beta]_t = [I] + [S][\beta] \quad (4\text{-}18)$$

Using (4-18), one may rewrite (4-15) in the form:

$$[S]\eta_o H_c] + [I]E_c] = -\frac{1}{\epsilon_o}[\Gamma]_t P] \quad (4\text{-}19)$$

Substituting (4-13) in (4-19), one obtains:

$$[T]E_c] = -[\Gamma]_t \frac{1}{\epsilon_o} P] \quad (4\text{-}20)$$

where we define:

$$[T] = [I] + [S][S] = \begin{bmatrix} 1-q^2-S_2^2 & S_1S_2 & S_1q \\ S_1S_2 & 1-q^2-S_1^2 & S_2q \\ S_1q & S_2q & 1-S_1^2-S_2^2 \end{bmatrix} \quad (4\text{-}21)$$

The displacement of the electron in the plasma from its equilibrium position is given in the present as follows:

$$\bar{R} = \bar{R}_1 e^{i(\omega t - k_oS_1x - k_oS_2y - k_oq z)} \quad (4\text{-}22)$$

where $\bar{R}_1$ = constant. Substituting (4-22) in (3-12) and using (4-14), one finds the total derivative or the material derivative for the present case in the form:

$$\frac{d\bar{R}}{dt} = \frac{D\bar{R}}{Dt} = i\omega[1 - \beta_xS_1 - \beta_yS_2 - \beta_zq]\bar{R} = i\omega s \bar{R} \quad (4\text{-}23)$$

where $s=1-\beta_xS_1 - \beta_yS_2 - \beta_zq$. Using (4-23) in (3-8), one obtains the constitutive relation (3-19) for the present case in the same form, provided that the above expression for s will be used. Equation (3-19) could be rewritten in a matrix form:

$$[I]E_c] + [\beta]\eta_o H_c] = \frac{1}{\epsilon_o}[Y]P] \quad (4\text{-}24)$$

where the constitutive relation matrix [Y] is given by:

$$[Y] = -\frac{s}{X}\begin{bmatrix} U & -iY_z & iY_y \\ iY_z & U & -iY_x \\ -iY_y & iY_x & U \end{bmatrix} \tag{4-25}$$

Equations (4-24) and (4-25) agree with the results given by Budden [1961] for no drift velocity $\bar{\beta} = 0$ and s = 1, except for the signs of $\bar{Y}$, which are opposite to ours [see note below equation (3-18)].

Substituting (4-13) in (4-24) and using definition (4-17), one obtains:

$$[\Gamma]\, E_c] = [Y]\, \frac{1}{\epsilon_0}\, P] \tag{4-26}$$

Equations (4-20) and (4-26) represent six homogeneous linear equations with six unknowns, and for a non-trivial solution the determinant of the coefficients should be zero:

$$\text{Det.} \begin{vmatrix} [\Gamma] & -[Y] \\ [T] & [\Gamma]_t \end{vmatrix} = 0 \tag{4-27}$$

where (4-27) is a 6 x 6 determinantal equation for q. By substituting (4-17), (4-21), and (4-25), one may find (4-27) in explicit form as in the original paper [Unz, 1965a].

From (4-20) and (4-26) one may obtain alternatively a set of three homogeneous linear equations with three unknowns by eliminating P]. Multiplying both sides of (4-20) by $[\Gamma]_t^{-1}$, the inverse of $[\Gamma]_t$ matrix, and substituting in (4-26), one obtains:

$$\left\{[\Gamma] + [Y]\,[\Gamma]_t^{-1}\,[T]\right\} E_c] = 0 \tag{4-28}$$

One requires for a non-trivial solution:

$$\text{Det.}\ \left|\,[\Gamma] + [Y]\,[\Gamma]_t^{-1}\,[T]\,\right| = 0 \tag{4-29}$$

where (4-29) is a 3 x 3 determinantal equation for q. Similarly one could derive from (4-20) and (4-26) the following 3 x 3 determinantal equations:

$$\text{Det.} \quad |[T] + [\Gamma]_t [Y]^{-1} [\Gamma]| = 0 \tag{4-30}$$

$$\text{Det.} \quad |[\Gamma]_t + [T][\Gamma]^{-1} [Y]| = 0 \tag{4-31}$$

$$\text{Det.} \quad |[Y] + [\Gamma][T]^{-1} [\Gamma]_t| = 0 \tag{4-32}$$

where the 3 x 3 determinantal equations (4-29) - (4-32) represent different approaches for the reduction of the 6 x 6 determinantal equation in (4-27), and all will give identical results for the factor q. In the next section it will be shown, that by applying the Minkowski theory for moving media to the convection current model, an equation identical to (4-29) will be obtained.

It has been shown [Unz, 1965a] that for the particular case of no drift velocity $\bar{\beta} = 0$, equation (4-30) will give the Booker quartic equation [Budden, 1961]. For the particular case of normal incidence $\theta_i = 0$, one has:

$$S_1 = S_2 = 0; \; \theta = 0; \; n = q; \; s = 1 - n\beta_L \tag{4-33}$$

For the normal incidence case it may be shown [Unz, 1965a] that (4-30) will give equation (3-66). For the normal incidence case it is also found that (4-29) will reduce to (3-68), which was derived from (3-66).

4.3 Convection Current Model (CCM)

For the non-relativistic case, the small signal harmonic time varying fields, in a magneto-plasma moving with a small velocity, obey the following Maxwell-Minkowski equations [Tai, 1965a] for the convection current model (CCM):

$$\nabla \times \bar{E} = -i\omega\mu_o \bar{H} \tag{4-34}$$

$$\nabla \times \bar{H} = \bar{J} + i\omega\epsilon_o \bar{E} \tag{4-35}$$

where $\bar{E}$, $\bar{H}$ represent the electric and the magnetic fields according to the Maxwell-Minkowski formulation, as extended by Tai [1965a].

The corresponding constitutive relations are given in the form [Tai, 1965a]:

$$\bar{D} = \epsilon_o \bar{E} \; ; \; \bar{B} = \mu_o \bar{H} \tag{4-36}$$

$$\bar{\bar{r}}' \cdot [\bar{J} - \bar{v}_o \nabla \cdot (\epsilon_o \bar{E})] = \bar{E} + \mu_o \bar{v}_o \times \bar{H} \tag{4-37}$$

$$\text{dyadic } \bar{\bar{r}}' = \frac{i}{X'\omega'\epsilon_o} [(1-iZ')\bar{\bar{I}} + i\bar{\bar{Y}}'] \tag{4-38}$$

where $\bar{\bar{I}}$ is the unit dyadic and the plasma parameters are defined by:

$$X' = \frac{Ne^2}{m\omega'^2 \epsilon_o} = \left(\frac{\omega_p}{\omega'}\right)^2 \tag{4-39}$$

$$Z' = \nu/\omega' \tag{4-40}$$

$$\bar{\bar{Y}}' = \begin{bmatrix} 0 & -Y_z' & Y_y' \\ Y_z' & 0 & -Y_x' \\ -Y_y' & Y_x' & 0 \end{bmatrix} \tag{4-41}$$

$$\bar{Y}' = Y_x' \hat{x} + Y_y' \hat{y} + Y_z' \hat{z} = \mu_o e \bar{H}_o / m\omega' \tag{4-42}$$

the primes refer to the rest system of reference embedded in the moving magneto-plasma.

Considering again the direction cosines of the incident plane electromagnetic wave to be S_1, S_2, and C, and that of the transmitted wave to be S_1, S_2, and q, the field components in the plasma will therefore contain factor:

$$e^{i(\omega t - k_o \bar{n} \cdot \bar{R})} = e^{i(\omega t - k_o S_1 x - k_o S_2 y - k_o qz)} \tag{4-43}$$

where $|\bar{n}| = n$ is the wave refractive index in the plasma, k_o is the free-space wave number, with the corresponding relationships to be found in (4-1) - (4-8).

Using (4-43) in (4-34) and (4-35), one obtains:

$$-ik_o(S_1\hat{x} + S_2\hat{y} + q\hat{z}) \times \bar{E} = -i\omega\mu_o \bar{H} \tag{4-44}$$

$$-ik_o(S_1\hat{x} + S_2\hat{y} + q\hat{z}) \times \bar{H} = \bar{J} + i\omega\epsilon_o\bar{E} \tag{4-45}$$

Using (4-44) and (4-45), one has:

$$\bar{J} = -i\omega\epsilon_o [(1 - n^2)\bar{E} + (S_1\hat{x} + S_2\hat{y} + q\hat{z})(S_1E_x + S_2E_y + qE_z)] \tag{4-46}$$

where $n^2 = S_1^2 + S_2^2 + q^2$.

Substituting for $\overline{J}$ in (4-37), one obtains:

$$-i\omega\epsilon_o \overline{\overline{r}}' \cdot [(1-n^2)\overline{E} + (S_1 E_x + S_2 E_y + qE_z)(S_1\hat{x} + S_2\hat{y} + q\hat{z} - \overline{\beta})] = \overline{E} + \overline{\beta} \times [(S_1\hat{x} + S_2\hat{y} + q\hat{z}) \times \overline{E}] \quad (4\text{-}47)$$

where $\overline{\beta} = \overline{v}_o/c$.

Taking the non-relativistic Doppler effect on the frequency as in (2-26) to be:

$$\omega' = \omega(1 - \overline{\beta} \cdot \overline{n}) = s\omega \quad (4\text{-}48)$$

where $s = 1 - \overline{\beta} \cdot \overline{n} = 1 - \beta_x S_1 - \beta_y S_2 - \beta_z q$, one obtains from (4-39) - (4-42):

$$X' = \frac{X}{s^2} \; ; \; Z' = \frac{Z}{s} \; ; \overline{Y}' = \frac{\overline{Y}}{s} \quad (4\text{-}49)$$

and (4-38) becomes

$$-i\omega\epsilon_o \overline{\overline{r}}' = \frac{1}{X}[(s - iZ)\overline{\overline{I}} + i\overline{\overline{Y}}]. \quad (4\text{-}50)$$

Using (4-50) in (4-47), one obtains:

$$[(s-iZ)\overline{\overline{I}} + i\overline{\overline{Y}}] \cdot [(n^2-1)\overline{E} - (S_1 E_x + S_2 E_y + qE_z)(S_1\hat{x} + S_2\hat{y} + q\hat{z} - \overline{\beta})] + X\overline{E} + X\overline{\beta} \times [(S_1\hat{x} + S_2\hat{y} + q\hat{z}) \times \overline{E}] = 0. \quad (4\text{-}51)$$

One can rewrite (4-51) in the form:

$$\overline{\overline{A}} \cdot \overline{E} = 0 \quad (4\text{-}52)$$

where the elements of the dyadic $\overline{\overline{A}}$ are:

$$A_{xx} = (n^2 - 1)U - (S_1 - \beta_x)S_1 U + iY_z(S_2 - \beta_y)S_1 - iY_y(q - \beta_z)S_1 + X(1 - S_2\beta_y - q\beta_z)$$

$$A_{xy} = -(S_1 - \beta_x)S_2 U - iY_z(n^2-1) + iY_z(S_2 - \beta_y)S_2 - iY_y(q - \beta_z)S_2 + X\beta_y S_1$$

$$A_{xz} = -(S_1 - \beta_x)qU + iY_z(S_2 - \beta_y)q + iY_y(n^2-1) - iY_y(q - \beta_z)q +$$

$$+ X\,\beta_z S_1$$

$$A_{yx} = iY_z(n^2 - 1) - iY_z(S_1 - \beta_x)S_1 - U(S_2 - \beta_y)S_1 + iY_x(q - \beta_z)S_1 +$$
$$+ X\,\beta_x S_2$$

$$A_{yy} = -iY_z(S_1 - \beta_x)S_2 + U(n^2 - 1) - U(S_2 - \beta_y)S_2 + iY_x(q - \beta_z)S_2 +$$
$$+ X(1 - \beta_x S_1 - \beta_z q)$$

$$A_{yz} = -iY_z(S_1 - \beta_x)q - U(S_2 - \beta_y)q - iY_x(n^2 - 1) + iY_x(q - \beta_z)q +$$
$$+ X\,\beta_z S_2$$

$$A_{zx} = -iY_y(n^2 - 1) + iY_y(S_1 - \beta_x)S_1 - iY_x(S_2 - \beta_y)S_1 - U(q - \beta_z)S_1 +$$
$$+ X\,\beta_x q$$

$$A_{zy} = iY_y(S_1 - \beta_x)S_2 + iY_x(n^2 - 1) - iY_x(S_2 - \beta_y)S_2 - U(q - \beta_z)S_2 +$$
$$+ X\,\beta_y q$$

$$A_{zz} = iY_y(S_1 - \beta_x)q - iY_x(S_2 - \beta_y)q + U(n^2 - 1) - U(q - \beta_z)q +$$
$$+ X(1 - S_1\beta_x - S_2\beta_y)$$

Taking $U = s - iZ = 1 - iZ - S_1\beta_x - S_2\beta_y - q\beta_z$.

For the particular case of normal incidence ($S_1 = S_2 = 0$, $q = n$), the components of $\bar{\bar{A}}$ agree with previous results [Tai, 1965a]. By setting the determinant of the matrix $\bar{\bar{A}}$ equal to zero, one obtains the characteristic equation for q. The determinantal equation Det. $|\bar{\bar{A}}| = 0$ thus obtained is found to be identical with (4-29), obtained previously by the polarization current model. It is found that the algebraic refractive index equation found for the general case of oblique incidence is of the eighth order in q, the same order as found previously in (3-68) for the particular case of normal incidence.

4.4 The Relativistic Refractive Index Equation

Let a laboratory system of coordinates S be stationary and be described by the space-time coordinates (x, y, z, t). Let another system of coordinates S' be moving with a constant velocity $\bar{v}_0$ with respect to the system S and be described by the space time coordinates (x', y', z', t'). For a plasma moving with velocity $\bar{v}_0$, the S' coordinate system moving with it will be called the rest system of the plasma coordinates. An oblique electromagnetic plane wave propagating in the moving plasma will have the following form with respect to the rest system S' of the plasma coordinates:

$$e^{-i(k_o'\bar{n}'\cdot\bar{r}' - \omega't')} = e^{-i[k_o'(S_1'x' + S_2'y' + q'z') - \omega't']} \tag{4-53}$$

where $k_o' = \omega'\sqrt{\mu_o\epsilon_o} = \omega'/c$. Since the oblique plane electromagnetic wave (4-53) propagates in a non-moving plasma in the rest system S', its q' factor has to obey the Booker quartic equation [Budden, 1961], which may be written in the following alternative compact form [Unz, 1966c]:

$$[U'(q'^2-C'^2)+X']\left\{[U'(q'^2-C'^2)+X'](U'-X')-Y'^2(q'^2-C'^2)\right\}+$$
$$+ X'(q'^2-C'^2)(S_1'Y_x' + S_2'Y_y' + q'Y_z')^2 = 0 \tag{4-54}$$

where we define $U' = 1 - iZ'$ and $C' = \cos\theta_i'$, θ_i' being the angle of incidence in the rest system S'. Using (4-1) and (4-4), one obtains:

$$n'^2 - q'^2 = S_1'^2 + S_2'^2 = 1 - C'^2 \tag{4-55}$$

from which one has:

$$q'^2 - C'^2 = n'^2 - 1 \tag{4-56}$$

Equation (4-54) was derived directly from the Appleton-Hartree equation [Budden, 1961], by using tensor analysis methods and the rotation of coordinates [Unz, 1966c].

The oblique plane electromagnetic wave in the moving plasma in the stationary laboratory coordinate system S will be of the form:

$$e^{-i(k_o\bar{n}\cdot\bar{r}-\omega t)} = e^{-i[k_o(S_1x + S_2y + qz) - \omega t]} \tag{4-57}$$

where $k_o = \omega\sqrt{\mu_o\epsilon_o} = \omega/c$, and similarly to (4-56):

$$q^2 - C^2 = n^2 - 1 \tag{4-58}$$

The scalar product of two four-vectors is invariant to rotation in space-time coordinates. As a result, it may be shown that the following quantities are invariant under the Lorentz transformation [Møller, 1952] in accordance with the results in chapter two:

$$x'^2 + y'^2 + z'^2 - c^2t'^2 = x^2 + y^2 + z^2 - c^2t^2 \tag{4-59}$$

$$k_x'^2 + k_y'^2 + k_z'^2 - \frac{\omega'^2}{c^2} = k_x^2 + k_y^2 + k_z^2 - \frac{\omega^2}{c^2} \tag{4-60}$$

$$k_x' x' + k_y' y' + k_z' z' - \omega' t' = k_x x + k_y y + k_z z - \omega t \tag{4-61}$$

Taking in (4-53) $\bar{k}' = k_o'\bar{n}'$ and in (4-57) $\bar{k} = k_o\bar{n}$, one has:

$$k_x' = k_o'S_1' \; ; \; k_y' = k_o'S_2' \; ; \; k_z' = k_o'q' \tag{4-62}$$

$$k_x = k_oS_1; \; k_y = k_oS_2; \; k_z = k_oq \tag{4-63}$$

Using (4-62) and (4-63) in (4-61), one finds that the phase variation of the plane waves in (4-53) and in (4-57) remains invariant under the Lorentz transformation [Møller, 1952; Papas, 1965] as in (2-24) for anisotropic and dispersive homogeneous magneto-plasma.

The refractive index equation for the q' factor in the moving rest system of coordinates S', with respect to which the plasma is stationary, is given by the Booker quartic equation (4-54). We would like now to derive the refractive index equation for the q factor in the laboratory system of coordinates S, with respect to which the plasma is moving with a constant velocity v_o in an arbitrary direction. This will be accomplished by using the Lorentz transformations and the plasma parameters transformations in (4-54).

The Lorentz transformation of the signal frequency is given by (2-26) in the form:

$$\omega' = \gamma(\omega - \bar{v}_o \cdot \bar{k}) \tag{4-64}$$

where $\gamma = 1/\sqrt{1-\beta^2} = 1/\sqrt{1-(v_o/c)^2}$. Defining, as above, $\bar{k} = k_o\bar{n} = \omega\bar{n}/c$ and $\bar{\beta} = \bar{v}_o/c$, one has from (4-64):

$$\omega' = \gamma\omega(1 - \bar{\beta}\cdot\bar{n}) = \gamma s \omega \tag{4-65}$$

where $\bar{\beta}(\beta_x, \beta_y, \beta_z)$, $\bar{n}(S_1, S_2, q)$, and $s = 1 - \bar{\beta}\cdot\bar{n} = 1 - \beta_x S_1 - \beta_y S_2 - \beta_z q$. Using (4-65), one has:

$$k_o' = \frac{\omega'}{c} = \frac{\gamma s \omega}{c} = \gamma s k_o \tag{4-66}$$

Using (4-62) and (4-63) in (4-60), one has:

$$k_o'^2(S_1'^2 + S_2'^2 + q'^2 - 1) = k_o^2(S_1^2 + S_2^2 + q^2 - 1) \tag{4-67}$$

Substituting (4-4), (4-56), and (4-58) in (4-67), one obtains:

$$k_o'^2(n'^2 - 1) = k_o'^2(q'^2 - C'^2) = k_o^2(q^2 - C^2) = k_o^2(n^2 - 1) \tag{4-68}$$

using (4-66) in (4-68), one obtains:

$$q'^2 - C'^2 = \frac{1}{\gamma^2 s^2}(q^2 - C^2) \tag{4-69}$$

It was shown in (2-33) that the plasma frequency is invariant under the Lorentz transformation:

$$\omega_p' = \omega_p \tag{4-70}$$

Using (4-65), (4-70), and the definitions [Budden, 1961] $X' = (\omega_p'/\omega')^2$ and $X = (\omega_p/\omega)^2$, one obtains:

$$X' = \frac{\omega_p'^2}{\omega'^2} = \frac{\omega_p^2}{\gamma^2 s^2 \omega^2} = \frac{X}{\gamma^2 s^2} \tag{4-71}$$

where $\frac{1}{\gamma^2} = 1 - \beta^2$ and $s = 1 - \bar{\beta}\cdot\bar{n}$.

It was shown in (2-36) that the collision frequency is transformed under the Lorentz transformation by:

$$\nu' = \gamma\nu \tag{4-72}$$

Using (4-65), (4-72), and the definitions [Budden, 1961] $Z' = \nu'/\omega'$ and $Z = \nu/\omega$, one obtains:

$$Z' = \frac{\nu'}{\omega'} = \frac{\gamma\nu}{\gamma s \omega} = \frac{\nu}{s\omega} = \frac{Z}{s} \tag{4-73}$$

$$U' = 1 - i Z' = 1 - i\frac{Z}{s} = \frac{1}{s}(s - i Z) \qquad (4\text{-}74)$$

It was shown in (2-39) that the gyro-frequency is transformed under the Lorentz transformation by:

$$\bar{\omega}_H' = \gamma^2 \bar{\omega}_H + \gamma(1 - \gamma)\frac{\bar{v}_o \cdot \bar{\omega}_H}{v_o^2}\bar{v}_o \qquad (4\text{-}75)$$

Using (4-65), (4-75), and the definitions [Budden, 1961] $\bar{Y}' = \bar{\omega}_H'/\omega'$, $\bar{Y} = \bar{\omega}_H/\omega$, one obtains:

$$\bar{Y}' = \frac{1}{s}\left[\gamma\bar{Y} + (1 - \gamma)\frac{\bar{v}_o \cdot \bar{Y}}{v_o^2}\bar{v}_o\right] \qquad (4\text{-}76)$$

From (4-76) one has:

$$Y'^2 = \bar{Y}' \cdot \bar{Y}' = \frac{1}{s^2}\left[\gamma^2 Y^2 + \frac{1 - \gamma^2}{v_o^2}(\bar{v}_o \cdot \bar{Y})^2\right] \qquad (4\text{-}77)$$

Taking $\bar{\beta} = \bar{v}_o/c$ and $\gamma^2 = 1/(1 - \beta^2)$, one has from (4-77):

$$Y'^2 = \frac{\gamma^2}{s^2}[Y^2 - (\bar{\beta} \cdot \bar{Y})^2] \qquad (4\text{-}78)$$

From (4-78) one obtains:

$$Y_L' = Y_L/s \quad \text{and} \quad Y_T' = \gamma Y_T/s$$

From (4-62) one has:

$$\bar{k}' \cdot \bar{Y}' = k_o'(S_1' Y_x' + S_2' Y_y' + q' Y_z') \qquad (4\text{-}79)$$

From (2-25) and (4-76) one has:

$$\bar{k}' \cdot \bar{Y}' = \frac{\gamma}{s}\left[\bar{k} \cdot \bar{Y} - \frac{\omega}{c^2}(\bar{v}_o \cdot \bar{Y})\right] \qquad (4\text{-}80)$$

Substituting (4-79) in (4-80) and using (4-66) and $\bar{k} = k_o\bar{n}$, $\bar{\beta} = \bar{v}_o/c$, one obtains after rearranging:

$$S_1' Y_x' + S_2' Y_y' + q' Y_z' = \frac{1}{s^2}\left[(\bar{n} \cdot \bar{Y}) - (\bar{\beta} \cdot \bar{Y})\right] \qquad (4\text{-}81)$$

Substituting the transformation relations (4-69), (4-71), (4-74), (4-78), and (4-81) into the Booker quartic equation (4-54), one obtains [Unz, 1968]:

$$[(s - iZ)(q^2 - C^2) + Xs] \cdot$$
$$\cdot \Big\{[(s - iZ)(q^2 - C^2) + Xs](s^2 - iZ\,s - \frac{X}{\gamma^2}) -$$
$$- \gamma^2 s(q^2 - C^2)[Y^2 - (\bar{\beta} \cdot \bar{Y})^2]\Big\} +$$
$$+ X(q^2 - C^2)\,[(\bar{n} \cdot \bar{Y}) - (\bar{\beta} \cdot \bar{Y})]^2 = 0 \qquad (4\text{-}82)$$

where X, $\bar{Y}$, Z are the drifting plasma parameters with respect to the laboratory stationary coordinate system S, and we define:

$$s = 1 - \bar{\beta} \cdot \bar{n} = 1 - \beta_x S_1 - \beta_y S_2 - \beta_z q$$
$$\bar{\beta} \cdot \bar{Y} = \beta_x Y_x + \beta_y Y_y + \beta_z Y_z$$
$$\bar{n} \cdot \bar{Y} = S_1 Y_x + S_2 Y_y + q Y_z$$

Equation (4-82) represents the relativistic refractive index equation for the q factor for the general case of a moving magneto-plasma, and it applies for large drift velocities of the plasma. It is an algebraic equation of the eighth order in q. Equation (4-82) was derived originally by Unz [1968].

Developing the non-relativistic determinantal equation Det$|\bar{A}| = 0$ in (4-52) and neglecting the second order relative velocity terms $\beta^2 \ll 1$, one will obtain, after rearranging, an eighth order algebraic equation for q. By developing the relativistic algebraic equation (4-82) and neglecting the second order relative velocity terms $\beta^2 \ll 1$, and $\gamma^2 \cong 1$ one will obtain, after rearranging, an equivalent eighth order algebraic equation for q. The two non-relativistic eighth-order algebraic equations for q could be shown to be identical after some tedious, cumbersome algebraic manipulations. The two equations for the particular case of normal incidence will be shown to be identical in the next section.

For the particular case of no plasma movement $\bar{\beta} = 0$, one has:

$$\bar{\beta} = 0; \quad \gamma = 1; \quad s = 1; \quad s - iZ = U \qquad (4\text{-}83)$$

and (4-82) will reduce to the Booker quartic equation (4-54) in its compact form [Unz, 1966c].

For the particular case of no static magnetic field $\overline{Y} = 0$ and no collisions $Z = 0$, the relativistic equation (4-82) will give the following two equations:

$$(1-\beta_x S_1-\beta_y S_2-\beta_z q)^2 - X(1-\beta_x^2-\beta_y^2-\beta_z^2) = 0 \qquad (4\text{-}84)$$

$$q^2 = C^2 - X = C^2 - \frac{\omega_p^2}{\omega^2} \qquad (4\text{-}85)$$

Of particular interest is the relativistic equation (4-85) for moving, isotropic plasma, which is identical to the equation for the non-moving, stationary, isotropic plasma [Budden, 1961]. The refractive index equation (4-85) is covariant under a plasma motion, and always has the same form with respect to all the inertial coordinate systems, which are at a constant uniform motion with respect to each other and with respect to the isotropic plasma.

4.5 The Relativistic Normal Incidence Case

For the particular case of normal incidence of the electromagnetic wave in the magneto-plasma, moving in arbitrary direction, one has:

$$\overline{n} = n\hat{z};\ q = n;\ S_1 = S_2 = 0;\ C = 1 \qquad (4\text{-}86)$$

From (4-86) one also has:

$$S = 1 - n\beta_z = 1 - n\beta_L;\ \overline{n}\cdot\overline{Y} = nY_L \qquad (4\text{-}87)$$

$$\overline{\beta}\cdot\overline{Y} = \beta_L Y_L + \overline{\beta}_T\cdot\overline{Y}_T \qquad (4\text{-}88)$$

Substituting (4-86) and (4-87) in (4-82), one has:

$$[(s-iZ)(n^2-1)+Xs]^2\left(s^2-iZs-\frac{X}{\gamma^2}\right) -$$
$$-\gamma^2 s(n^2-1)[(s-iZ)(n^2-1)+Xs][Y^2-(\overline{\beta}\cdot\overline{Y})^2] +$$
$$+X(n^2-1)[nY_L-(\overline{\beta}\cdot\overline{Y})]^2 = 0 \qquad (4\text{-}89)$$

Equation (4-89) represents the relativistic refractive index equation for the case of normal incidence on a magneto-plasma, moving with relativistic velocities

$v_0 < c$.

For the particular case in which the plasma is moving in the longitudinal direction $\hat{z}$ of the static magnetic field, and the electromagnetic wave propagates in the same longitudinal direction, one has for the longitudinal propagation case:

$$\bar{n} = n\hat{z},\ \bar{Y} = Y_L\hat{z};\ \bar{\beta} = \beta_L\hat{z} \tag{4-90}$$

where $\hat{z}$ is the unit vector in the longitudinal direction. From (4-90) one has:

$$\gamma^2[Y^2 - (\bar{\beta}\cdot\bar{Y})^2] = \gamma^2 Y_L^2(1-\beta_L^2) = Y_L^2 \tag{4-91}$$

$$[nY_L - \bar{\beta}\cdot\bar{Y}]^2 = Y_L^2(n-\beta_L)^2 =$$

$$= Y_L^2[(n^2-1)(1-\beta_L^2) + (1-n\beta_L)^2] = Y_L^2\left[\frac{n^2-1}{\gamma^2} + s^2\right] \tag{4-92}$$

Substituting (4-91) and (4-92) into (4-89), one obtains after rearranging:

$$\left[s^2 - iZs - \frac{X}{\gamma^2}\right]\left\{[(s-iZ)(n^2-1) + Xs]^2 - (n^2-1)^2 Y_L^2\right\} = 0 \tag{4-93}$$

Taking $s = 1 - n\beta_L$ and solving (4-93), one has:

$$(1-n\beta_L)^2 - iZ(1-n\beta_L) - X(1-\beta_L^2) = 0 \tag{4-94}$$

$$(n^2-1)(1-n\beta_L - iZ \pm Y_L) + (1-n\beta_L)X = 0 \tag{4-95}$$

Equations (4-94) and (4-95) represent the relativistic refractive index equations for magneto-plasma moving in the longitudinal direction. The relativistic refractive index equation (4-95) was found previously [Unz, 1966a].

For the particular case of no static magnetic field $\bar{Y} = 0$ and no collisions $Z = 0$, the relativistic equation (4-89) will give, by taking $Y_L = 0$ and $Z = 0$ in (4-94) and (4-95):

$$(1-n\beta_L)^2 - X(1-\beta_L^2) = 0 \tag{4-96}$$

$$n^2 = 1 - X = 1 - \frac{\omega_p^2}{\omega^2} \qquad (4\text{-}97)$$

The relativistic equation (4-97) for moving plasma is identical with the equation in non-moving, stationary isotropic plasma [Budden, 1961].

For the particular case of small plasma velocities $v_o^2 \ll c^2$ or $\beta_L^2 \ll 1$ (4-94) becomes:

$$(1 - n\beta_L)^2 - iZ(1 - n\beta_L) - X = 0 \qquad (4\text{-}98)$$

Equation (4-98) is identical with the one derived previously in (3-33) for the non-relativistic case.

The relativistic refractive index equation (4-95) is identical with the corresponding refractive index equation for the non-relativistic case found in (3-38). In other words, the refractive index equation (4-95) for the longitudinal propagation case is the same for both relativistic large plasma velocities and for non-relativistic small plasma velocities. This fact was pointed out originally elsewhere [Unz, 1966a].

For the case of small plasma drift velocity $\beta^2 \ll 1$ one will have in (4-89) for the first order approximation:

$$\frac{1}{\gamma^2} = 1 - \beta^2 \cong 1; \; \gamma^2 \cong 1 \qquad (4\text{-}99)$$

$$Y^2 - (\bar{\beta} \cdot \bar{Y})^2 = Y^2(1 - \beta^2\cos^2\alpha) \cong Y^2 \qquad (4\text{-}100)$$

where α is the angle between the vectors $\bar{\beta}$ and $\bar{Y}$, and one also has:

$$[nY_L - (\bar{\beta} \cdot \bar{Y})]^2 \cong n^2 Y_L^2 - 2nY_L(\bar{\beta} \cdot \bar{Y}) =$$
$$= n^2 Y_L^2 - 2nY_L[\beta_L Y_L + \bar{\beta}_T \cdot \bar{Y}_T] \qquad (4\text{-}101)$$

where $(\bar{\beta} \cdot \bar{Y})^2$ has been neglected in accordance with (4-100). Substituting (4-99) - (4-101) in (4-89), one has:

$$[(s - iZ)(n^2 - 1) + Xs]^2(s^2 - iZs - X) -$$

$$- sY^2(n^2 - 1)[(s - iZ)(n^2 - 1) + Xs] +$$

$$+ X(n^2 - 1)[n^2Y_L^2 - 2n\beta_L Y_L^2 - 2nY_L(\bar{\beta}_T \cdot \bar{Y}_T)] = 0 \quad (4\text{-}102)$$

Taking $s = 1 - n\beta_L$ and $Y^2 = Y_L^2 + Y_T^2$ in (4-102), rearranging and neglecting $\beta_L^2 \ll 1$, one obtains:

$$[(n^2 - 1)(s - iZ) + Xs]^2(s^2 - iZs - X) -$$

$$-Y_L^2 (n^2 - 1)^2 (s^2 - iZs - X) -$$

$$-Y_T^2 s(n^2 - 1)[(n^2 - 1)(s - iZ) + Xs] -$$

$$-2XY_L n(n^2 - 1)(\bar{\beta}_T \cdot \bar{Y}_T) = 0 \quad (4\text{-}103)$$

Equation (4-103) represents the non-relativistic refractive index equation for the case of normal incidence in a plasma moving with small velocity $v_o \ll c$, and it has been found as a particular case from the relativistic equation (4-89). Equation (4-103) is identical with (3-70), found by using the non-relativistic magneto-ionic theory for a moving temperate plasma.

V

OBLIQUE WAVE PROPAGATION IN MOVING WARM MAGNETO-PLASMA

5.1 Introduction

In chapter three and chapter four the refractive index equation for moving, temperate magneto-plasmas was considered, where the effects of the plasma temperature were neglected. In the present chapter the plasma is regarded to be warm, and the temperature and pressure are taken into account, by considering the linearized equations of plasma dynamics. Two approaches for the derivation of the refractive index characteristic equations are given. The first approach, which is non-relativistic and applicable for non-relativistic velocities only, is based on the single-fluid continuum theory, described by Oster [1960], and was considered by Chawla, Rao, and Unz [1966]. The refractive index characteristic equation found by using this approach is in a determinantal equation form. The expansion of this determinantal equation to an algebraic form is rather cumbersome. The type of boundary value problems to which this determinantal refractive index equation applies are the ones where the boundary is stationary with respect to the laboratory observer, while the plasma is moving with respect to it, the continuum being maintained through the mechanism of ionization at the boundary The second approach, which is relativistic, used the method of plasma-parameters transformations. This method was first correctly used by Unz [1966a] and was extended to the present case by Chawla and Unz [1969a]. Two types of boundary value problems are considered using this approach. One of them is described above, and the other corresponds to the plsama together with its boundary, moving with respect to a laboratory observer. The refractive index equations using this approach are found in an algebraic equations form.

The plasma is cosidered to be neutral, uniform, and in equilibrium. The entire plasma, consisting of electrons, ions, and neutral particles, is drifting with a finite uniform velocity $\overline{v}_0$ with respect to a laboratory observer. It is further assumed that the electron gas is warm and compressible, and that there exists a static magnetic field $\overline{H}_0$. The direction of propagation of the waves in the warm plasma, the drift velocity of the plasma, and the direction of the static magnetic field are assumed to be arbitrary.

5.2 The Non-Relativistic Refractive Index Equation

In this section the cartesian tensor notation [Holt and Haskell, 1965] is used. The small signal theory is used and the equations of the conservation of the mass and the conservation of momentum are linearized. The Maxwell equations of electrodynamics are given in the form:

$$\epsilon_{ijk}E_{k,j} = -\mu_0\dot{H}_i \tag{5-1}$$

$$\epsilon_{ijk}H_{k,j} = \epsilon_0\dot{E}_i + J_i \tag{5-2}$$

where ϵ_{ijk} is the Levi-Civita symbol defined by:

$$\epsilon_{ijk} = \begin{cases} 0 \text{ if any two indices are same,} \\ +1 \text{ if indices are in cyclic order,} \\ -1 \text{ if indices are in noncyclic order.} \end{cases}$$

The comma among the indices indicates a partial derivative with respect to the coordinate index following it, and the dot atop the variables represents a partial time derivative.

E_i and H_i are the electromagnetic field quantities, J_i is the current density, where no electromagnetic sources are assumed, ϵ_0 and μ_0 being respectively the permittivity and the permeability of free space. The linearized conservation of mass equation, or the continuity equation, can be written as:

$$J_{i,i} = -\dot{\rho} \tag{5-3}$$

where one has:

$$\rho = -eN \tag{5-4}$$

$$J_i = -eN_o v_i - eNv_{oi} \tag{5-5}$$

The subscript o denotes the stationary values of the variables of the plasma, (-e) is the electron charge, v_{oi} is the plasma drift velocity, N is the perturbed electron number density, and v_i is the perturbed velocity of the electrons. In (5-5) the perturbation of the ions is assumed to be zero, and the gross convection current term $J_{oi} = eN({}^{i}v_{oi} - {}^{e}v_{oi})$ is neglected [Oster, 1960], since the ion and electron drift velocities are assumed to be the same.

The linearized conservation of momentum equation for the electron gas may be written in the form:

$$mN_o(\dot{v}_i + v_{oj}v_{i,j} + \nu v_i) + mN\nu v_{oi} =$$

$$= -p_{,i} - eN_oE_i - e\mu_o\epsilon_{ijk}(N_ov_{oj}H_k + N_ov_jH_{ok} + Nv_{oj}H_{ok}) \tag{5-6}$$

where H_{ok} is the static magnetic field, ν is the phenomenological collision frequency, m is the electron mass, and p is the perturbed pressure of the electron gas. The gas is considered to be an inviscid, perfect fluid. The equation of state of the electron gas may be expressed in the form [Oster, 1960; Unz, 1966d]:

$$P_{,i} = \alpha KT_oN_{,i} \tag{5-7}$$

where T_o is the equilibrium temperature of the plasma, K is the Boltzmann constant, and α is the ratio of the electron gas specific heat at constant pressure to its specific heat at a constant volume. The electron gas is considered to obey the adiabatic law, and in essence it is equivalent to the truncation of tne higher moments of the Boltzmann equation by neglecting the heat flux tensor, and assuming a scalar pressure [Holt and Haskell, 1965].

From equations (5-2) and (5-3) one may obtain the divergence equation:

$$\epsilon_oE_{\ell,\ell} = -eN. \tag{5-8}$$

Equations (5-1) - (5-7) are the basic equations

for the plasma under consideration, and (5-8) is a useful auxiliary equation.

Eliminating H_i in (5-1) and (5-2), one obtains:

$$E_{\ell,\ell i} - E_{i,\ell\ell} = -\mu_o \epsilon_o \ddot{E}_i - \mu_o \dot{J}_i, \tag{5-9}$$

Substituting (5-5) in (5-9) and rearranging, one has:

$$eN_o \dot{v}_i = \frac{1}{\mu_o}[E_{\ell,\ell i} - E_{i,\ell\ell} + \mu_o \epsilon_o \ddot{E}_i] - e\dot{N}v_{oi} \tag{5-10}$$

where the two dots atop E_i denote a second partial time derivative. Substituting (5-8) in (5-10), one obtains an equation for the jth component:

$$eN_o \dot{v}_j = \frac{1}{\mu_o}[E_{\ell,\ell j} - E_{j,\ell\ell} + \mu_o \epsilon_o \ddot{E}_j] + \epsilon_o \dot{E}_{\ell,\ell} v_{oj}. \tag{5-11}$$

Equation (5-11) gives the perturbed velocity in terms of the electric field. Taking the gradient of (5-8) and substituting in (5-7), it follows that:

$$p_{,i} = -(\alpha KT_o \epsilon_o / e) E_{\ell,\ell i}. \tag{5-12}$$

Taking the partial time derivative of (5-6) and using (5-12), one obtains after rearranging:

$$N_o \left[\ddot{v}_i + v_{oj}\dot{v}_{i,j} + \nu\dot{v}_i + \frac{e\mu_o}{m}\epsilon_{ijk}\dot{v}_j H_{ok}\right] + \dot{N}\left[\nu v_{oi} + \frac{e\mu_o}{m}\epsilon_{ijk} v_{oj} H_{ok}\right] =$$

$$= \frac{\alpha KT_o \epsilon_o}{me}\dot{E}_{\ell,\ell i} - \frac{eN_o}{m}\dot{E}_i - \frac{e\mu_o N_o}{m}\epsilon_{ijk} v_{oj}\dot{H}_k \tag{5-13}$$

Using (5-1) and (5-8) in (5-13), one obtains

$$N_o Q_{ij}\dot{v}_j - \frac{\epsilon_o}{e}\dot{E}_{\ell,\ell} R_{ij} v_{oj} = \frac{\alpha KT_o \epsilon_o}{me}\dot{E}_{\ell,\ell i} - \frac{eN_o}{m}\dot{E}_i +$$

$$+ \frac{eN_o}{m}(E_{j,i} - E_{i,j}) v_{oj} \tag{5-14}$$

where the following identities [Holt and Haskell, 1965] have been used:

$$\epsilon_{ijk} = \epsilon_{kij}; \quad \epsilon_{kij}\epsilon_{k\ell m} = \delta_{i\ell}\delta_{jm} - \delta_{im}\delta_{j\ell} \tag{5-15}$$

and δ_{ij} is the Kronecker delta.

The tensor operators Q_{ij} and R_{ij} are defined by:

$$Q_{ij} \equiv \begin{bmatrix} \frac{\partial}{\partial t} + v_{0j}\frac{\partial}{\partial x_j} + \nu & \frac{e\mu_o}{m} H_{03} & -\frac{e\mu_o}{m} H_{02} \\ -\frac{e\mu_o}{m} H_{03} & \frac{\partial}{\partial t} + v_{0j}\frac{\partial}{\partial x_j} + \nu & \frac{e\mu_o}{m} H_{01} \\ \frac{e\mu_o}{m} H_{02} & -\frac{e\mu_o}{m} H_{01} & \frac{\partial}{\partial t} + v_{0j}\frac{\partial}{\partial x_j} + \nu \end{bmatrix}$$

$$R_{ij} \equiv \begin{bmatrix} \nu & \frac{e\mu_o}{m} H_{03} & -\frac{e\mu_o}{m} H_{02} \\ -\frac{e\mu_o}{m} H_{03} & \nu & -\frac{e\mu_o}{m} H_{o1} \\ \frac{e\mu_o}{m} H_{02} & -\frac{e\mu_o}{m} H_{01} & \nu \end{bmatrix}$$

Substituting (5-11) in (5-14) and rearranging, one obtains:

$$Q_{ij}[E_{\ell,\ell j} - E_{j,\ell\ell} + \epsilon_o\mu_o\ddot{E}_j] + \frac{1}{c^2}(Q_{ij} - R_{ij})\dot{E}_{\ell,\ell}v_{oj} - \frac{a^2}{c^2}\dot{E}_{\ell,\ell i} + \frac{\mu_o e^2 N_o}{m}\dot{E}_i - \frac{\mu_o e^2 N_o}{m}(E_{j,i} - E_{i,j})v_{oj} = 0 \quad (5\text{-}16)$$

where $a^2 = \alpha KT_o/m$, a = electron acoustic velocity, $c^2 = 1/\epsilon_o\mu_o$, and c = velocity of light in free space.

Equation (5-16) is the general electric field equation for homogeneous, gyrotropic, warm, drifting plasma. Equation (5-16) may be solved for E_i, which can be used in turn to find other electrodynamic and fluid dynamic variables.

For the particular case of no drift velocity and harmonic time variation $e^{+i\omega t}$, the tensor operator Q_{ij} can be written in the form:

$$Q_{ij} = i\omega U\,\delta_{ij} + i \cdot i\omega Y_{ij}$$

$$= i\omega \begin{bmatrix} U & 0 & 0 \\ 0 & U & 0 \\ 0 & 0 & U \end{bmatrix} + i \cdot i\omega \begin{bmatrix} 0 & -Y_3 & Y_2 \\ Y_3 & 0 & -Y_1 \\ -Y_2 & Y_1 & 0 \end{bmatrix} \qquad (5\text{-}17)$$

where $U = 1 - iZ$, $Z = \nu/\omega$, and $Y_i = e\mu_o H_{0i}/m\omega$. Substituting (5-17) in (5-16) for $v_{0j} \equiv 0$, one obtains after rearranging in a vector form

$$-(1/k_o^2)\nabla x \nabla x\overline{E} + (1/Uk_1^2)\nabla(\nabla \cdot \overline{E}) + (1 - X/U)\overline{E} +$$

$$+ (1/Uk_o^2)(\nabla x \nabla x\overline{E} - k_o^2\overline{E}) \text{ x } i\overline{Y} = 0 \qquad (5\text{-}18)$$

where

$$X = e^2N_o/\omega^2\epsilon_o m,$$

$$k_1 = \omega/a = (c/a)k_o = \text{acoustic wave number},$$

$$k_o = \omega/c = \text{free-space wave number}.$$

Equation (5-18) agrees with a previous result [Unz, 1966b].

Let us now consider an oblique plane plasma wave transmitted in the general magneto-plasma. Let the field variations contain the factor $\exp[i(\omega t - k_o\overline{n} \cdot \overline{r})]$, where

$$\overline{n} = S_1\hat{x} + S_2\hat{y} + q\hat{z},$$

$$n = |\overline{n}| = \text{wave refractive index in the plasma},$$

$$n^2 = S_1^2 + S_2^2 + q^2.$$

Substituting the field variation $\exp[i(\omega t - k_o\overline{n} \cdot \overline{r})]$ in (5-16), one obtains the following equation in a vector form:

$$U(1 - n^2)\overline{E} + (U - a^2/c^2)(\overline{n} \cdot \overline{E})\overline{n} - [(1 - n^2)\overline{E} + (\overline{n} \cdot \overline{E})\overline{n}] \text{ x } i\overline{Y} -$$

$$-(1 - \overline{n} \cdot \overline{\beta})(\overline{n} \cdot \overline{E})\overline{\beta} - X[\overline{E} + \overline{\beta} \text{ x } (\overline{n} \text{ x } \overline{E})] = 0, \qquad (5\text{-}19)$$

where U is now defined for the moving plasma in the form:

$$U = s - iZ; \quad s = 1 - \beta_1 S_1 - \beta_2 S_2 - \beta_3 q;$$

$$\overline{\beta} = \overline{v}_o/c = (\beta_1, \beta_2, \beta_3); \quad \overline{Y} = (Y_1, Y_2, Y_3).$$

Equation (5-19) can be rearranged and written in the form:

$$\begin{bmatrix} A_{11} & A_{12} & A_{13} \\ A_{21} & A_{22} & A_{23} \\ A_{31} & A_{32} & A_{33} \end{bmatrix} \begin{bmatrix} E_1 \\ E_2 \\ E_3 \end{bmatrix} = 0 \tag{5-20}$$

where the coefficients of the matrix $\overline{\overline{A}}$ are:

$$A_{11} = U(1 - n^2) + (U - a^2/c^2)S_1^2 + iS_1(qY_2 - S_2Y_3) - s\beta_1 S_1 - $$
$$-X(1 - \beta_2 S_2 - \beta_3 q),$$

$$A_{12} = (U - a^2/c^2)S_1S_2 + iS_2qY_2 - i(1 - S_1^2 - q^2)Y_3 - s\beta_1 S_2 - X\beta_2 S_1,$$

$$A_{13} = (U - a^2/c^2)S_1 q + i(1 - S_1^2 - S_2^2)Y_2 - iS_2qY_3 - s\beta_1 q - X\beta_3 S_1,$$

$$A_{21} = (U - a^2/c^2)S_1S_2 - iS_1qY_1 + i(1 - S_2^2 - q^2)Y_3 - s\beta_2 S_1 - X\beta_1 S_2,$$

$$A_{22} = U(1 - n^2) + (U - a^2/c^2)S_2^2 - iS_2(qY_1 - S_1Y_3) - s\beta_2 S_2 -$$
$$-X(1 - \beta_1 S_1 - \beta_3 q),$$

$$A_{23} = (U - a^2/c^2)S_2 q - i(1 - S_1^2 - S_2^2)Y_1 + iS_1qY_3 - s\beta_2 q - X\beta_3 S_2,$$

$$A_{31} = (U - a^2/c^2)S_1 q + iS_1S_2Y_1 - i(1 - S_2^2 - q^2)Y_2 - s\beta_3 S_1 - X\beta_1 q,$$

$$A_{32} = (U - a^2/c^2)S_2 q + i(1 - S_1^2 - q^2)Y_1 - iS_1S_2Y_2 - s\beta_3 S_2 - X\beta_2 q,$$

$$A_{33} = U(1 - n^2) + (U - a^2/c^2)q^2 + iq(S_2Y_1 - S_1Y_2) - s\beta_3 q -$$
$$-X(1 - \beta_1 S_1 - \beta_2 S_2).$$

The refractive index equation is found by equating the determinant of the matrix [A] to zero. The variable in this equation is really q, with S_1 and

S_2 assumed to be known, but since n may be expressed in terms of q, we may call this equation in q the refractive index equation. For the particular case of no drift, the matrix [A] reduces to the matrix [F] given by Unz [1966b], where the determinant of [F] is found to be of the 6th order in q. The refractive index equation in the present case of the 8th order in q.

For the particular case of temperate plasma, with all other properties of the medium and the wave remaining the same, one takes a = 0 in (5-19) and obtains:

$$U(1-n^2)\bar{E}+U(\bar{n}\cdot\bar{E})\bar{n}-[(1-n^2)\bar{E}+(\bar{n}\cdot\bar{E})\bar{n}]\times i\bar{Y}-$$
$$-(1-\bar{n}\cdot\bar{\beta})(\bar{n}\cdot\bar{E})\bar{\beta}-X[\bar{E}+\bar{\beta}\times(\bar{n}\times\bar{E})]=0 \quad (5\text{-}21)$$

For comparison, consider the equation (4-51) derived in chapter four for the temperate plasma. After some rearrangements, one has (4-51) in the following form:

$$U(1-n^2)\bar{E}+U(\bar{n}\cdot\bar{E})\bar{n}-[(1-n^2)\bar{E}+(\bar{n}\cdot\bar{E})(\bar{n}-\bar{\beta})]\times i\bar{Y}-$$
$$-(1-\bar{n}\cdot\bar{\beta}-iZ)(\bar{n}\cdot\bar{E})\bar{\beta}-X[\bar{E}+\bar{\beta}\times(\bar{n}\times\bar{E})]=0 \quad (5\text{-}22)$$

The extra terms present in (5-22) as compared to (5-21) are due to the fact that in the formulation of the momentum conservation equation for the temperate case, one neglects the term containing the tensor R_{ij}, which is associated with the warm plasma only. Due to the presence of the tensor R_{ij}, some terms are operated by $(Q_{ij}-R_{ij})$ rather than Q_{ij}. This gives rise to the cancellation of some terms present in (5-22), and one obtains the form (5-21). This has been pointed out by Unz [1966d] for the isotropic case. Therefore, the formulation for two different types of plasma, with regard to their temperature, must be different. The conservation of momentum equation considered in the temperate anisotropic plasma is given by

$$mN_o(\dot{v}_i+v_{oj}v_{i,j}+\nu v_i)=-eN_oE_i-e\mu_o\epsilon_{ijk}(N_ov_{oj}H_k+N_ov_jH_{ok}), \quad (5\text{-}23)$$

where (5-23) corresponds to (5-6) in the present section [Unz, 1966d].

5.3 The Relativistic Refractive Index Equation

The present section applies to two types of boundary value problems. The first type corresponds to the stationary boundary, which is termed as the non-drifting boundary problem. The second type corresponds to the moving boundary, which is termed as the drifting boundary problem. The algebraic refractive index characteristic equation for the wave refractive index observed in the laboratory system of coordinates is found for the two cases, by first applying the coordinate rotation transformations, and then the Lorentz transformations, to the refractive index equation found in the plasma rest system for waves propagating normal to the boundary. It is found that one obtains an eighth-order algebraic equation for the non-drifting boundary value problem, and a sixth-order algebraic equation for the drifting boundary value problem. The results obtained in this section apply to all plasma drift velocities $v_o < c$.

Let us denote the rest system of the plasma by S' and the laboratory system by S. All the primed quantities refer to S', and the unprimed quantities to S. The non-drifting boundary is assumed to be located at $z = 0$, and the drifting boundary at $z' = 0$, with the semi-infinite plasma being contained in the regions $z > 0$ or $z' > 0$, respectively, for the two cases. An electromagnetic wave is assumed to be incident from the free space $z < 0$ or $z' < 0$, respectively, at an arbitrary angle to the plasma-free space interface.

We shall first derive the refractive index equation for an oblique wave, with respect to the boundary, in the rest system of the plasma in a suitable form, by applying the rotation of coordinates [Unz, 1966c]. To do so, we shall denote the double primed system (x'', y'', z'') as the system of coordintaes in which the oblique wave in the plasma propagates along the $+z''$ axis, and the single primed system (x', y', z') as the system in which z' axis is normal to the boundary. Both systems of coordinates in this section refer to the rest system of the plasma with the time variable t' and both are embedded in the moving plasma.

The refractive index equation for a wave propagating along the $+z''$ direction, in a stationary, warm, general magneto-plasma, using the linearized, single fluid, continuum theory of plasma dynamics, may be found in a determinantal form [Unz, 1966b]:

$$\begin{vmatrix} F''_{11} & F''_{12} & F''_{13} \\ F''_{21} & F''_{22} & F''_{23} \\ F''_{31} & F''_{32} & F''_{33} \end{vmatrix} = 0 \qquad (5\text{-}24)$$

where

$$F''_{11} = U'(1 - n''^2 - \frac{X'}{U'})$$

$$F''_{12} = -i(1 - n''^2)Y_{z''}$$

$$F''_{13} = i\,Y_{y''}$$

$$F''_{21} = i(1 - n''^2)Y_{z''}$$

$$F''_{22} = U'(1 - n''^2 - \frac{X'}{U'})$$

$$F''_{23} = -i\,Y_{x''}$$

$$F''_{31} = -i(1 - n''^2)Y_{y''}$$

$$F''_{32} = i(1 - n''^2)Y_{x''}$$

$$F''_{33} = U'(1 - n''^2 - \frac{X'}{U'}) + (U' - \delta')n''^2$$

The wave is assumed to be of the form $e^{i(\omega' t' - k'_0 \bar{n}'' \cdot \bar{R}'')}$, where $\bar{n}'' = n''\hat{z}''$, $S''_1 = S''_2 = 0$, n'' being the wave refractive index, $k'_0 = \omega'/c$, c being the velocity of light in free space, $\delta' = a'^2/c^2$, a' being the acoustic velocity in the electron gas, $U' = 1 - iZ'$, $Z' = \nu'/\omega'$, ν' being the mean collision frequency, $(Y_{x''}, Y_{y''}, Y_{z''}) = \bar{Y}'' = e\bar{B}''_0/m'\omega'$, $\bar{B}''_0$ being the static magnetic field, (-e) is the charge and m' is the mass of an electron, $X' = \omega_p'^2/\omega'^2$, and ω'_p is the plasma frequency. Expanding the determinant on the left side of (5-24), one may obtain [Unz, 1966c]:

$$U'^2(1 - n''^2 - \frac{X'}{U'})^2[(U'-X') - \delta' n''^2] -$$

$$-U'(1 - n''^2 - \frac{X'}{U'}) \cdot (1 - n''^2)Y''^2 +$$

$$(1 - n''^2)[-X' + \delta'(1 - n''^2)]n''^2 Y_{z''}^2 = 0 \qquad (5\text{-}25)$$

where $Y'' = |\bar{Y}''| = \sqrt{Y_{x''}^2 + Y_{y''}^2 + Y_{z''}^2}$,

Using a method suggested elsewhere [Unz, 1966c], let us rotate the coordinate system (x'',y'', z'') into a new system (x',y',z'), keeping the vectors $\bar{n}''$ and $\bar{Y}''$ intact. Since the dot product remains invariant under the rotation of coordinates [Aris, 1962], one has:

$$\bar{n}'' \cdot \bar{R}'' = \bar{n}' \cdot \bar{R}' \qquad (5\text{-}26)$$

$$n''^2 = n'^2 = S_1'^2 + S_2'^2 + q'^2;\ \bar{n}' = S_1'\hat{x} + S_2'\hat{y} + q'\hat{z} \qquad (5\text{-}27)$$

$$Y''^2 = Y'^2 = Y_{x'}^2 + Y_{y'}^2 + Y_{z'}^2;\ \bar{Y}' = \frac{e\bar{B}'o}{m'\omega'} \qquad (5\text{-}28)$$

$$\bar{n}'' \cdot \bar{Y}'' = \bar{n}' \cdot \bar{Y}' = n''Y_{z''} = S_1'Y_{x'} + S_2'Y_{y'} + q'Y_{z'} \qquad (5\text{-}29)$$

$$1 - n''^2 = 1 - S_1'^2 - S_2'^2 - q'^2 = C'^2 - q'^2 \qquad (5\text{-}30)$$

where $C'^2 = 1 - S_1'^2 - S_2'^2$, and the incident electromagnetic wave in free space has the exponential variation $e^{i[\omega't' - k_0'(S_1'x' + S_2'y' + q'z')]}$. The transmitted wave in the plasma will have the exponential variation $e^{i[\omega't' - k_0'(S_1'x + S_2'y + q'z)]}$.

Substituting the transformations (5-26) - (5-30) in (5-25), one may obtain after rearrangement of the terms:

$$U'(C'^2 - q'^2 - \frac{X'}{U'})[U'(C'^2 - q'^2 - \frac{X'}{U'})(U' - X') -$$

$$- \delta'U'(C'^2 - q'^2 - \frac{X'}{U'})(1 - C'^2 + q'^2) - (C'^2 - q'^2)Y'^2] +$$

$$+[-X'(C'^2 - q'^2) + \delta'(C'^2 - q'^2)^2] \cdot (S_1'Y_{x'} + S_2'Y_{y'} + q'Y_{z'})^2 = 0 \qquad (5\text{-}31)$$

For temperate plasma ($\delta' = 0$), (5-31) reduces to an alternative form of the Booker quartic equation [Budden, 1961] as derived by Unz [1966c]. Equation (5-31) may be rewritten in the following form:

$$\alpha_6'q'^6 + \alpha_5'q'^5 + \alpha_4'q'^4 + \alpha_3'q'^3 + \alpha_2'q'^2 + \alpha_1'q' + \alpha_0' = 0 \qquad (5\text{-}32)$$

where we define:

$$\alpha_6' = -\delta'(U'^2 - Y_{z'}^2)$$

$$\alpha_5' = 2\delta' Y_{z'}(S_1' Y_{x'} + S_2' Y_{y'})$$

$$\alpha_4' = [U'^2(U' - X') - U'Y'^2 + X'Y_{z'}^2] + \delta'[(S_1'Y_{x'} + S_2'Y_{y'})^2 + 2U'^2(C'^2 - \frac{X'}{U'}) - U'^2(1 - C'^2) - 2C'^2 Y_{z'}^2]$$

$$\alpha_3' = 2Y_{z'}(S_1'Y_{x'} + S_2'Y_{y'})(X' - 2\delta' C'^2)$$

$$\alpha_2' = -2(C'^2U' - X')(U' - X')U' + 2(C'^2U' - X')Y'^2 - X'C'^2Y_{z'}^2 + X'(S_1'Y_{x'} + S_2'Y_{y'})^2 + X'Y'^2 + \delta'(C'^2U' - X')[2U'(1 - C'^2) - (C'^2U' - X')] + \delta' C'^4 Y_{z'}^2 - 2\delta' C'^2(S_1'Y_{x'} + S_2'Y_{y'})^2$$

$$\alpha_1' = -2C'^2 Y_{z'}(X' - \delta' C'^2)(S_1'Y_{x'} + S_2'Y_{y'})$$

$$\alpha_0' = (C'^2U' - X')[(C'^2U' - X')(U' - X') - C'^2Y'^2] - C'^2(X' - \delta' C'^2)(S_1'Y_{x'} + S_2'Y_{y'})^2 - \delta'(1 - C'^2)(C'^2U' - X')^2$$

Equation (5-32) is identical with the one obtained by Unz [1966b]. We shall, however, use equation (5-31) in the next two sections, since this form facilitates the use of the Lorentz transformations similarly to section 4.4. It should be noted that equations (5-31) and (5-32) are in terms of q'. Assuming S_1' and S_2' to be known, one can then find the refractive index n' using (5-27), (5-30), and the knowledge of q'.

5.4 Non-Drifting Boundary Problem

In order to obtain the refractive index characteristic equation in the laboratory system of coordinates S, we need to know the transformation of all the quantities appearing in (5-32).

Consider the following plasma parameters:

$$X' = \frac{\omega_p'^2}{\omega'^2}; \quad \bar{Y}' = \frac{\bar{\omega}_H'}{\omega'}; \quad U' = 1 - iZ'; \quad Z' = \frac{\nu'}{\omega'}; \quad \delta' = \frac{a'^2}{c^2} \qquad (5\text{-}33)$$

It has been shown [Unz, 1966a; Chawla and Unz, 1966b]

in section 2.4 that:

$$\omega_p' = \omega_p; \quad \nu' = \gamma\nu; \quad a' = \gamma a \tag{5-34}$$

where $\gamma = (1 - \beta^2)^{-1/2}$, $\beta = \frac{v_0}{c}$, $\bar{v}_0$ being the drift velocity of the plasma.

Assuming that the plasma is neutral, and that there is no net electrostatic field in the laboratory coordinate system S, one has [Papas, 1965] from (2-38) and (2-39):

$$\bar{B}_0' = \gamma\bar{B}_0 + (1 - \gamma)\frac{\bar{v}_0 \cdot \bar{B}_0}{v_0^2}\bar{v}_0 \tag{5-35}$$

$$\bar{\omega}_H' = \frac{e\bar{B}_0'}{m'} = \gamma^2\bar{\omega}_H + \gamma(1 - \gamma)\frac{\bar{v}_0 \cdot \bar{\omega}_H}{v_0^2}\bar{v}_0 \tag{5-36}$$

where $m' = m/\gamma$ and $\bar{\omega}_H$ is the gyrofrequency in the laboratory stationary coordinate system.

The transformation of the wave frequency is given by (4-65):

$$\omega' = \gamma\omega s \tag{5-37}$$

where $s = 1 - \bar{\beta}\cdot\bar{n} = (1 - \beta_1 S_1 - \beta_2 S_2 - \beta_3 q)$ and $\bar{k} = k_0(S_1\hat{x} + S_2\hat{y} + q\hat{z})$.

Substituting (5-34) - (5-37) in (5-33), one obtains, as in section 4.4:

$$X' = \frac{X}{\gamma^2 s^2}, \quad \bar{Y}' = \frac{1}{s}[\gamma\bar{Y} + (1-\gamma)\frac{\bar{v}_0 \cdot \bar{Y}}{v_0^2}\bar{v}_0],$$

$$Z' = Z/s, \quad U' = (s - iZ)/s, \quad \delta' = \gamma^2\delta. \tag{5-38}$$

From the invariance of the dot product under Lorentz transformation, one may obtain, as in (4-60):

$$-k_0^2 + k_x^2 + k_y^2 + k_z^2 = -k_0'^2 + k_x'^2 + k_y'^2 + k_z'^2 \tag{5-39}$$

where $k_0^2 = \omega^2/c^2$, $k_0'^2 = \omega'^2/c^2$.

Equation (5-39) may be reduced [Unz, 1968], as in (4-68):

$$k_0^2(n^2 - 1) = k_0^2(q^2 - C^2) = k_0'^2(q'^2 - C'^2) = k_0'^2(n'^2 - 1) \tag{5-40}$$

Since $k_o = \frac{\omega}{c}$ and $k_o' = \frac{\omega'}{c}$, using (5-37) in (5-40), one obtains, as in (4-69):

$$(q'^2 - C'^2) = \frac{1}{\gamma^2 s^2}(q^2 - C^2) \tag{5-41}$$

From (5-38) one has, as in (4-78) and (4-81):

$$Y'^2 = \bar{Y}' \cdot \bar{Y}' = \frac{\gamma^2}{s^2}[Y^2 - (\bar{\beta} \cdot \bar{Y})^2] \tag{5-42}$$

$$(S_1' Y_x' + S_2' Y_y' + q' Y_z') = \frac{1}{s^2}[(\bar{n} \cdot \bar{Y}) - (\bar{\beta} \cdot \bar{Y})] \tag{5-43}$$

Substituting (5-38) and (5-41) - (5-43) in (5-31), multiplying the resulting equation by $\gamma^4 s^8$, and rearranging, one obtains:

$$\begin{aligned} &[(s - iZ)(q^2 - C^2) + sX]^2 \cdot \\ &\cdot [(s^2 - iZs - \frac{X}{\gamma^2}) - \delta(\gamma^2 s^2 + q^2 - C^2)] - \\ &- [(s - iZ)(q^2 - C^2) + sX] \cdot \gamma^2 s (q^2 - C^2)[Y^2 - (\bar{\beta} \cdot \bar{Y})^2] + \\ &+ [X(q^2 - C^2) + \gamma^2 \delta (q^2 - C^2)^2] \cdot [(\bar{n} \cdot \bar{Y}) - (\bar{\beta} \cdot \bar{Y})]^2 = 0 \end{aligned} \tag{5-44}$$

For the particular case of temperate plasma one substitutes $\delta = 0$ in (5-44), which subsequently reduces to:

$$\begin{aligned} &[(s-iZ)(q^2 - C^2) + sX]\{[(s-iZ)(q^2-C^2) + sX](s^2-iZs-\frac{X}{\gamma^2}) - \\ &-\gamma^2 s(q^2-C^2)[Y^2-(\bar{\beta} \cdot \bar{Y})^2]\} + X(q^2-C^2)[(\bar{n} \cdot \bar{Y}) - (\bar{\beta} \cdot \bar{Y})]^2 = 0 \end{aligned} \tag{5-45}$$

Equation (5-45) is identical with (4-82), derived originally by Unz [1968].

Equation (5-44) may be written in the following form:

$$\sum_{n=0}^{8} \alpha_n q^n = 0 \tag{5-46}$$

where we have:

$$\alpha_8 = a_6 b_2$$

$$\alpha_7 = a_6 b_1 + a_5 b_2$$

$$\alpha_6 = a_6 b_0 + a_5 b_1 + a_4 b_2 + c_3 d_3 + e_4 f_2$$

$$\alpha_5 = a_5 b_0 + a_4 b_1 + a_3 b_2 + c_3 d_2 + c_2 d_3 + e_4 f_1$$

$$\alpha_4 = a_4 b_0 + a_3 b_1 + a_2 b_2 + c_3 d_1 + c_2 d_2 + c_1 d_3 + e_4 f_0 + e_2 f_2$$

$$\alpha_3 = a_3 b_0 + a_2 b_1 + a_1 b_2 + c_3 d_0 + c_2 d_1 + c_1 d_2 + c_0 d_3 + e_2 f_1$$

$$\alpha_2 = a_2 b_0 + a_1 b_1 + a_0 b_2 + c_2 d_0 + c_1 d_1 + c_0 d_2 + e_2 f_0 + e_0 f_2$$

$$\alpha_1 = a_1 b_0 + a_0 b_1 + c_1 d_0 + c_0 d_1 + e_0 f_1$$

$$\alpha_0 = a_0 b_0 + c_0 d_0 + e_0 f_0$$

and we define:

$$a_6 = \beta_3^2;\ a_5 = 2\beta_3(iZ - \chi);\ a_4 = 2\beta_3^2(X - C^2) + \chi^2 - 2iZ\chi - Z^2$$

$$a_3 = 2\beta_3[iZ(X - 2C^2) - 2\chi(X - C^2)]$$

$$a_2 = \beta_3^2(C^2 - X)^2 + 2\chi^2(X - C^2) - 2iZ\chi(X - 2C^2) + 2Z^2C^2$$

$$a_1 = 2\beta_3[iZC^2(C^2 - X) - \chi(C^2 - X)^2]$$

$$a_0 = \chi^2(C^2 - X)^2 - 2iZ\chi C^2(C^2 - X) - Z^2C^4$$

$$b_2 = (\beta_3^2 - \delta\gamma^2\beta_3^2 - \delta);\ b_1 = \beta_3(iZ - 2\chi + 2\chi\gamma^2\delta)$$

$$b_0 = (\chi^2 - iZ\chi - \frac{X}{\gamma^2} - \delta\gamma^2\chi^2 + \delta C^2)$$

$$c_3 = \beta_3;\ c_2 = (iZ - \chi);\ c_1 = \beta_3(X - C^2)$$

$$c_0 = \chi(C^2 - X) - iZC^2;\ d_3 = -\beta_3\tau;\ d_2 = \chi\tau;\ d_1 = \beta_3 C^2\tau$$

$$d_0 = -\chi C^2\tau$$

$$e_4 = \gamma^2\delta;\ e_2 = (X - 2\gamma^2\delta C^2);\ e_0 = C^2(C^2\delta\gamma^2 - X)$$

$$f_2 = Y_3^2;\ f_1 = 2Y_3(S_1Y_1 + S_2Y_2 - \overline{\beta}\cdot\overline{Y})$$

$$f_0 = (S_1Y_1 + S_2Y_2)^2 + (\bar{\beta}\cdot\bar{Y})^2 - 2(\bar{\beta}\cdot\bar{Y})(S_1Y_1 + S_2Y_2)$$

$$\chi = (1 - S_1\beta_1 - S_2\beta_2); \quad \tau = \gamma^2[Y^2 - (\bar{\beta}\cdot\bar{Y})^2]$$

Equation (5-46) is of the eighth order in q, thus representing eight characteristic waves in a warm, drifting magneto-plasma with a nondrifting boundary. For non-relativistic velocities ($v_0 \ll c$), one may substitute $\gamma = 1$ in (5-46), and the result then represents the expansion of the determinantal equation given in section 5.2. In the next section we shall determine the refractive index equation for warm magneto-plasma drifting together with its boundary.

5.5 Drifting Boundary Problem

In accordance with the relativistic analysis in section 6.3 for a plasma drifting together with its boundary, the boundary conditions require that the wave frequency ω' in the rest system of the plasma S' will be transformed [Chawla and Unz, 1967a] as follows:

$$\omega' = \gamma\omega(1 - S_1\beta_1 - S_2\beta_2 - C\beta_3) = \gamma\omega\zeta \qquad (5\text{-}47)$$

where $\zeta = 1 - S_1\beta_1 - S_2\beta_2 - C\beta_3$.

From (5-47) one has:

$$k_0' = \gamma k_0 \zeta \qquad (5\text{-}48)$$

Substituting (5-34) - (5-36) and (5-47) - (5-48) in (5-33), one obtains, similarly to (5-38):

$$X' = \frac{X}{\gamma^2\zeta^2}; \quad \bar{Y}' = \frac{1}{\zeta}\left[\gamma\bar{Y} + (1-\gamma)\frac{\bar{v}_0\cdot\bar{Y}}{v_0^2}\bar{v}_0\right];$$

$$Z' = \frac{Z}{\zeta}; \quad U' = \frac{1}{\zeta}(\zeta - iZ); \quad \delta' = \gamma^2\delta \qquad (5\text{-}49)$$

From (5-49) one has, similarly to (5-42) and (5-43):

$$Y'^2 = \bar{Y}'\cdot\bar{Y}' = \frac{\gamma^2}{\zeta^2}[Y^2 - (\bar{\beta}\cdot\bar{Y})^2] \qquad (5\text{-}50)$$

$$(S_1'Y_x' + S_2'Y_y' + q'Y_z') = \frac{1}{\zeta^2}[(\bar{n}\cdot\bar{Y}) - (\bar{\beta}\cdot\bar{Y})] \qquad (5\text{-}51)$$

From the invariance of the dot product under the Lorentz transformation, one obtains, similarly to (5-41):

$$(q'^2 - C'^2) = \frac{1}{\gamma^2\zeta^2}(q^2 - C^2) \tag{5-52}$$

Substituting (5-49) - (5-52) in (5-31), and multiplying the resulting equation by $\gamma^4\zeta^8$, one obtains:

$$\begin{aligned}
&[(\zeta-iZ)(q^2-C^2)+X\zeta]^2 \cdot [(\zeta^2-iZ\zeta-\frac{X}{\gamma^2}) - \delta(\zeta^2\gamma^2+q^2-C^2)]- \\
&-[(\zeta-iZ)(q^2-C^2)+X\zeta] \cdot \gamma^2\zeta(q^2-C^2)[Y^2-(\bar{\beta}\cdot\bar{Y})^2]+ \\
&+[X(q^2-C^2)+\gamma^2\delta(q^2-C^2)^2] \cdot [(\bar{n}\cdot\bar{Y})-(\bar{\beta}\cdot\bar{Y})]^2 = 0
\end{aligned} \tag{5-53}$$

Equation (5-53) is an algebraic equation of the sixth order in q, thus representing six characteristic waves in a warm magneto-plasma, drifting together with its boundary. This should be expected, since all that belongs to the system is in motion [Lorentz, 1952]. Equation (5-53) may be written in the following form:

$$\sum_{n=1}^{6} \alpha_n q^n = 0 \tag{5-54}$$

where we have:

$$\alpha_6 = a_4b_2 + d_4e_2$$

$$\alpha_5 = d_4e_1$$

$$\alpha_4 = a_4b_0 + a_2b_2 + c_4 + d_4e_0 + d_2e_2$$

$$\alpha_3 = d_2e_1$$

$$\alpha_2 = a_2b_0 + a_0b_2 + c_2 + d_2e_0 + d_0e_2$$

$$\alpha_1 = d_0e_1$$

$$\alpha_0 = a_0b_0 + c_0 + d_0e_0$$

and we define:

$$a_4 = (\zeta - iZ)^2; \quad a_2 = 2[X\zeta - (\zeta - iZ)C^2](\zeta - iZ)$$

$$a_0 = [X\zeta - (\zeta - iZ)C^2]^2$$

$$b_2 = -\delta;\ b_0 = (\zeta^2 - iZ - \frac{X}{\gamma^2}) - \delta(\zeta^2\gamma^2 - C^2)$$

$$c_4 = \gamma^2\zeta(iZ - \zeta)[Y^2 - (\overline{\beta}\cdot\overline{Y})^2]$$

$$c_2 = \gamma^2\zeta[2C^2(\zeta - iZ) - X\zeta][Y^2 - (\overline{\beta}\cdot\overline{Y})^2]$$

$$c_0 = C^2\gamma^2\zeta[X\zeta - C^2(\zeta - iZ)][Y^2 - (\overline{\beta}\cdot\overline{Y})^2]$$

$$d_4 = \gamma^2\delta;\ d_2 = X - 2\gamma^2\delta C^2;\ d_0 = C^2(\gamma^2\delta C^2 - X)$$

$$e_2 = Y_3^2;\ e_1 = 2Y_3(S_1Y_1 + S_2Y_2 - \overline{\beta}\cdot\overline{Y});\ e_0 = (S_1Y_1 + S_2Y_2 - \overline{\beta}\cdot\overline{Y})^2$$

For the particular case of the plasma being temperate, one substitutes $\delta = 0$ in (5-54) and obtains a fourth order algebraic equation, thus predicting the four waves of the magneto-ionic theory, with the refractive indices modified due to the motion of the plasma with respect to a laboratory observer.

5.6 The Longitudinal Propagation Case

In the longitudinal propagation case, the constant drift velocity of the plasma, the imposed static magnetic field, and the direction of the electromagnetic wave propagation are all in the same longitudinal direction. Taking $\hat{z}$ as the longitudinal direction, one has for this case:

$$\overline{n} = n\hat{z};\ S_1 = 0;\ S_2 = 0;\ q = n;\ C = 1 \tag{5-55}$$

$$\overline{Y} = Y_L\hat{z};\ Y_1 = 0;\ Y_2 = 0;\ Y_3 = Y_z = Y_L \tag{5-56}$$

$$\overline{\beta} = \beta_L\hat{z};\ \beta_1 = 0;\ \beta_2 = 0;\ \beta_3 = \beta_z = \beta_L \tag{5-57}$$

Substituting (5-55) - (5-57) in (5-20) for the non-relativistic case, one obtains for the determinantal equation terms:

$$A_{11} = A_{22} = U(1 - n^2) - Xs = -[U(n^2 - 1) + Xs] \tag{5-58}$$

$$A_{33} = U(1 - n^2) + (U - \delta)n^2 - sn\beta_L - X \tag{5-59}$$

$$A_{21} = -A_{12} = +i(1 - n^2)Y_L \tag{5-60}$$

$$A_{13} = A_{31} = A_{23} = A_{32} = 0 \tag{5-61}$$

where $U = s - iZ$ and $s = 1 - n\beta_L$. The determinantal equation of the matrix (5-20) will thus become for the present case:

$$\begin{vmatrix} A_{11} & -A_{21} & 0 \\ A_{21} & A_{11} & 0 \\ 0 & 0 & A_{33} \end{vmatrix} = 0 \tag{5-62}$$

where (5-62) will give the equations:

$$A_{33} = 0; \; A_{11}^2 + A_{21}^2 = 0 \tag{5-63}$$

Using (5-58) - (5-60) in (5-63), one obtains the following equations after rearranging, where $\delta = a^2/c^2$:

$$(1 - n\beta_L)^2 - iZ - X - \delta n^2 = 0 \tag{5-64}$$

$$(n^2 - 1)(1 - n\beta_L - iZ \pm Y_L) + X(1 - n\beta_L) = 0 \tag{5-65}$$

Equation (5-64) is identical with the result found previously for propagation in isotropic warm plasma with no magneto-static field (Unz, 1966d). Equation (5-65) is identical with the relativistic result found in (4-95). Equations (5-64) and (5-65) represent the non-relativistic refractive index equations for the warm moving magneto-plasma for the longitudinal case.

The corresponding relativistic refractive index equations for the longitudinal propagation case may be found by using (5-55) and (5-57) as follows:

$$\gamma^2[Y^2 - (\bar{\beta} \cdot \bar{Y})^2] = \gamma^2(1 - \beta_L^2)Y_L^2 = Y_L^2 \tag{5-66}$$

$$[(\bar{n} \cdot \bar{Y}) - (\bar{\beta} \cdot \bar{Y})]^2 = (n - \beta_L)^2 Y_L^2 \tag{5-67}$$

Using the above in (5-44), one obtains:

$$[(s-iZ)(n^2-1) + Xs]^2 [(s^2 - iZs - \frac{X}{\gamma^2}) - \delta(\gamma^2 s^2 + n^2 - 1)] -$$

$$-[(s - iZ)(n^2 - 1) + Xs]s(n^2 - 1)Y_L^2 +$$

$$+[X(n^2-1)+\gamma^2\delta(n^2-1)^2](n-\beta_L)^2Y_L^2 = 0 \quad (5\text{-}68)$$

where $s = 1 - n\beta_L$ and (5-68) is the relativistic equation for the longitudinal propagation case.

It is found that:

$$(n-\beta_L)^2 = (1-n\beta_L)^2 + (n^2-1)(1-\beta_L^2) = s^2 + \frac{n^2-1}{\gamma^2} \quad (5\text{-}69)$$

Substituting (5-69) in (5-68) and combining the Y_L^2 terms, one has:

$$[(s-iZ)(n^2-1)+Xs]^2[(s^2-iZs-\frac{X}{\gamma^2}) - \delta(\gamma^2s^2+n^2-1)] -$$

$$-Y_L^2(n^2-1)^2[(s^2-iZs-\frac{X}{\gamma^2}) - \delta(\gamma^2s^2+n^2-1)] = 0 \quad (5\text{-}70)$$

From (5-70) one obtains:

$$[(s^2 - iZs - \frac{X}{\gamma^2}) - \delta(\gamma^2s^2+n^2-1)]\cdot$$

$$\cdot\{[(s-iZ)(n^2-1)+Xs]^2 - Y_L^2(n^2-1)^2\} = 0 \quad (5\text{-}71)$$

where $\delta = a^2/c^2$ and $s = 1 - n\beta_L$.

From (5-71) one obtains the following relativistic refractive index equations, using (5-69):

$$(1-n\beta_L)^2 - iZ(1-n\beta_L) - X(1-\beta_L^2) - \frac{\delta}{1-\beta_L^2}(n-\beta_L)^2 = 0 \quad (5\text{-}72)$$

$$(n^2-1)(1-n\beta_L - iZ \pm Y_L) + X(1-n\beta_L) = 0 \quad (5\text{-}73)$$

Equations (5-72) and (5-73) represent the relativistic refractive index equations. Equation (5-72) represents the longitudinal plasma wave mode of propagation, and equation (5-73) represents the transverse electromagnetic modes of propagation. The modes of propagation are independent of each other for the longitudinal propagation case. Equation (5-72) was found previously [Chawla and Unz, 1966b]. Equation (5-73) is identical with the result found previously in (5-65), (4-95), and (3-38). For the particular non-relativsitc case of small plasma drift velocity $\beta_L^2 \ll 1$, and taking $\beta_L \ll n$,, equation (5-72) becomes:

$$(1 - n\beta_L)^2 - iZ(1 - n\beta_L) - X - \delta n^2 = 0 \tag{5-74}$$

Equation (5-74) for the non-relativistic case is identical with (5-64) for Z = 0. However, equations (5-64) and (5-74) do not agree for the case of collisions $Z \neq 0$. The reason for this disagreement was discussed at the end of section 5.2 by comparing (5-6) with (5-23). A detailed discussion of the reason for disagreement between (5-64) and (5-74) may also be found elsewhere [Unz, 1966d], where both results were derived for the non-relativistic isotropic plasma case with no magnetostatic field. For the particular case of cold plasma $T_o = 0$, $a = 0$, and $\delta = 0$, equation (5-72) and (5-73) for the relativistic case reduce to (4-94) and (4-95), found previously.

VI

NORMAL INCIDENCE ON SEMI-INFINITE TEMPERATE MAGNETO-PLASMA MOVING THROUGH FREE SPACE

6.1 Introduction

It is well known [Sommerfeld, 1964b; Becker and Sauter, 1964] that when electromagnetic waves are reflected from a moving mirror, a change occurs in the frequency as well as in the amplitude of the wave. This effect is significant in the case when the velocity of the moving mirror is comparable to the velocity of light. The problem of reflection and transmission of electromagnetic waves by a moving semi-infinite dielectric has been considered by Tai [1964b] and Yeh [1965]. The results indicate that the change is significant when the velocity of the semi-infinite dielectric is comparable to the velocity of light in free space for the case in which the medium moves along the boundary, and when it is comparable to the velocity of light in the dielectric medium for the case in which the medium moves normal to the boundary. Under laboratory conditions, it is not possible to impart such high velocities to macroscopic bodies; however, the use of a relativistically moving electron beam, in order to enhance the above mentioned effects, was proposed by Landecker [1952].

Landecker [1952] gave the reflection coefficients for a circularly polarized electromagnetic wave, normally incident on a semi-infinite plasma ("electron beam"), moving along and across a uniform static magnetic field. Since a circularly polarized wave excites a transmitted wave of the same circular polarization, the polarization of the reflected wave does not change. The analysis in this case is simplified considerably, being parallel to the analysis for the isotropic case. The difference between the two cases is that in the case of the longitudinal anisotropic plasma, for the two possible polarizations of the incident wave, there are two possible refractive indices of the transmitted wave, and therefore two results for the reflection coefficients for the longitudinal magnetic field. Reflection and transmission of electromagnetic waves by a moving plasma, in the absence of static magnetic field, has been recently considered by Yeh [1966].

In the present chapter we shall evaluate the reflection and the transmission coefficients for a linearly polarized electromagnetic wave, normally incident on a semi-infinite plasma, moving along a uniform static magnetic field in the longitudinal direction, normal to the boundary. The present problem is more complicated since two circularly polarized waves, both right-hand and left-hand circular polarizations, will be excited in the plasma, and hence the polarization of the reflected wave will be different from that of the incident wave. The formulas for the reflection coefficients will involve in this case the refractive indices of both the transmitted waves, as compared to only one refractive index for the case of a circularly polarized incident wave. The final results will be given in terms of the plasma parameters observed in the laboratory system of coordinates, using the transformations considered by Unz [1966a]. It should be pointed out that Landecker [1952] gave his results in terms of the parameters of the plasma observed in the moving system of coordinates, namely, the rest system of the drifting plasma.

Two methods of solution for the evaluation of the reflection and the transmission coefficients will be presented. The first, which is the non-relativistic solution [Unz, 1967a], treats the problem completely in the laboratory frame of reference, and is valid for non-relativistic velocities only. The second, which is the relativistic solution [Chawla and Unz, 1967a], treats the problem in the rest system of the plasma, and then Lorentz transformations are applied in order to obtain the results in the laboratory system. The second solution is valid also for relativistic velocities. It is found that for the present normal incidence case the two solutions give identical results [Chawla and Unz, 1968]. The numerical results for the presnet case [Chawla and Unz, 1968] will be discussed in section 6.5. Based on the solutions for the linearly polarized incident electromagnetic wave, the circularly polarized incident wave case will be discussed in section 6.6.

6.2 The Non-Relativistic Solution

Let a linearly polarized plane electromagnetic wave traveling in free space in the positive $\hat{z}$ direction be normally incident on the drifting boundary of a semi-infinite temperate magneto-plasma, which is drifting with a constant velocity $\underline{v}_0 = v_0\hat{\underline{z}}$, with an externally impressed static magnetic field $\underline{H} = H_0\hat{z}$. Let the incident wave be given by (Fig. 1):

$$\bar{E}_I = E_x^I \hat{x} e^{i(\omega_I t - k_I z)} \; ; \; \bar{H}_I = \frac{1}{\eta_0} E_x^I \hat{y} e^{i(\omega_I t - k_I z)} \tag{6-1}$$

where E_x^I is a constant, ω_I is the circular frequency of the incident wave, $k_I = \omega_I/c = \omega_I(\mu_0 \, \epsilon_0)^{1/2}$, $\eta_0 = (\mu_0/\epsilon_0)^{1/2}$ and μ_0, ϵ_0 are the free space permeability and permittivity, $\hat{x}$, $\hat{y}$, $\hat{z}$ being the unit vectors of the corresponding system x, y, z.

The drifting gyrotropic plasma boundary, assumed to be located at $z = v_0 t$, will produce in general a superposition of two reflected waves in free space. space:

$$\bar{E}_{R1} = E_x^R \hat{x} e^{i(\omega_R t + k_R z)} \; ; \; \bar{H}_{R1} = -\frac{1}{\eta_0} E_x^R \hat{y} e^{i(\omega_R t + k_R z)} \tag{6-2}$$

$$\bar{E}_{R2} = E_y^R \hat{y} e^{i(\omega_R t + k_R z)} \; ; \; \bar{H}_{R2} = \frac{1}{\eta_0} E_y^R \hat{x} e^{i(\omega_R t + k_R z)} \tag{6-3}$$

where E_x^R, E_y^R are constants, ω_R is the circular frequency of the reflected wave, and $k_R = \omega_R/c$.

In addition, one will have transmitted waves in the drifting magneto-plasma of the type $e^{i(\omega_T t - k_T z)}$. From Maxwell equations one obtains for the plasma waves [Unz, 1962] the following relations, as in (3-23) - (3-27):

$$H_x = -\frac{k_T}{\mu_0 \omega_T} E_y = -\frac{n}{\eta_0} E_y ; \; H_y = \frac{k_T}{\mu_0 \omega_T} E_x = \frac{n}{\eta_0} E_x \tag{6-4}$$

$$\epsilon_0(n^2 - 1)E_x = s \, P_x ; \quad \epsilon_0(n^2 - 1)E_y = s \, P_y \tag{6-5}$$

where $s = 1 - n\beta_L$, ω_T is the circular frequency of the

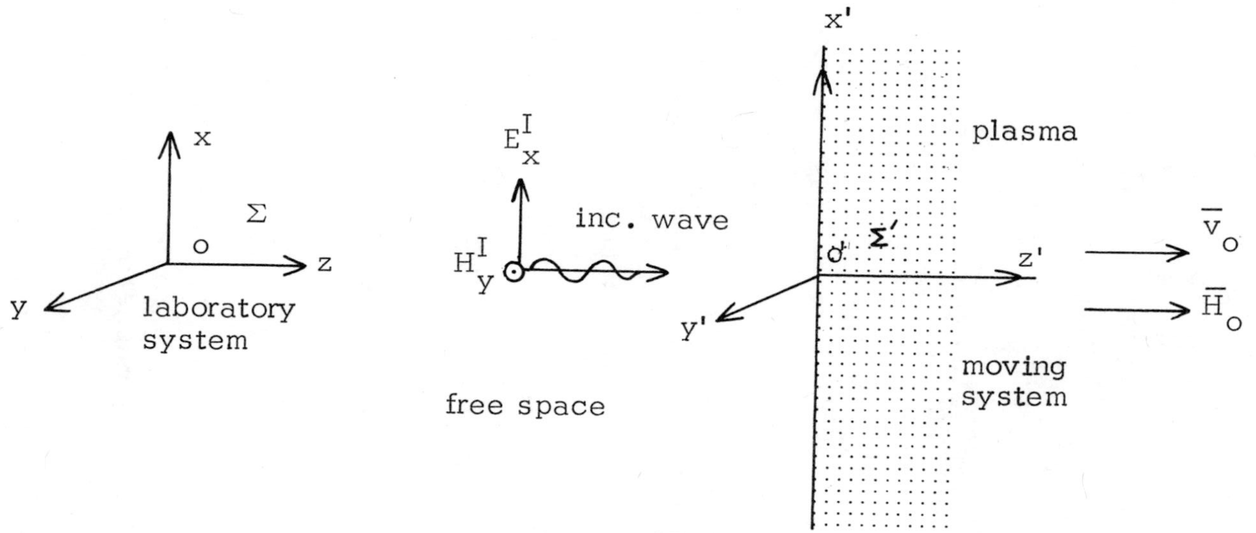

Figure 1. Normal Incidence on Moving Semi-Infinite Magneto-Plasma

transmitted electromagnetic plasma waves, k_T is the corresponding wave number, P_x, P_y are the space polarization components, $\beta_L = v_0/c$, $n = ck_T/\omega_T = c/u$ is the refractive index, and u is the phase velocity of the wave in the drifting magneto-plasma. From the constitutive relations, one has for the plasma waves [Unz, 1965a], as in (3-29) - (3-30):

$$\epsilon_0 X_T E_x = -U_T P_x + iY_T P_y \tag{6-6}$$

$$\epsilon_0 X_T E_y = -U_T P_y - iY_T P_x \tag{6-7}$$

where

$$X_T = \frac{\omega_p^2}{\omega_T^2}, \quad Y_T = \frac{\omega_{HL}}{\omega_T}, \quad Z_T = \frac{\nu}{\omega_T}, \quad U_T = s - iZ_T, \quad s = 1 - n\beta_L,$$

ω_p is the plasma frequency, ω_{HL} is the longitudinal gyromagnetic frequency, and ν is the phenomenological collision frequency of the plasma. Substituting (6-5) into (6-6) - (6-7), one obtains:

$$[U_T(n^2 - 1) + sX_T]E_x = +iY_T(n^2 - 1)E_y \tag{6-8}$$

$$[U_T(n^2 - 1) + sX_T]E_y = -iY_T(n^2 - 1)E_x \tag{6-9}$$

By equating the determinant of the homogeneous equations (6-8) - (6-9) to zero, one may obtain for a non-trivial solution:

$$(n^2 - 1)[1 - n\beta_L - iZ_T \pm Y_T] + (1 - n\beta_L)X_T = 0 \tag{6-10}$$

where (6-10) is the well known cubic refractive index equation [Unz, 1965a], found in (3-38) for the longitudinally drifting magneto-plasma, with respect to the circular frequency ω_T of the transmitted plasma waves. Using (6-10) in (6-8), and the result in (6-4) - (6-5), one obtains the polarization ρ of the waves in the plasma as follows:

$$\rho = \frac{E_y}{E_x} = \frac{P_y}{P_x} = -\frac{H_x}{H_y} = \mp i \tag{6-11}$$

where (6-11) is identical with the result of the classical magneto-ionic theory for non-moving stationary plasma [Budden, 1961].

The effective values of the electric field $\overline{E}_{eff}$ and magnetic field $\overline{H}_{eff}$ in a moving medium are given by [Fano, Chu, and Adler, 1960]:

$$\overline{E}_{eff} = \overline{E} + \overline{v}_o \times \mu_o \overline{H} \tag{6-12}$$

$$\overline{H}_{eff} = \overline{H} - \overline{v}_o \times \epsilon_o \overline{E} \tag{6-13}$$

It has been shown [Sommerfeld, 1964a] that the tangential boundary conditions at a moving boundary require:

$$\hat{n} \times [\overline{E}_{eff}]^{+}_{-} = 0; \quad \hat{n} \times [\overline{H}_{eff}]^{+}_{-} = 0 \tag{6-14}$$

where there is no surface current density on the boundary and $[F]^{+}_{-}$ represents the discontinuity $F^{+}-F^{-}$, with $\pm$ signs relative to $\hat{n}$, the unit vector normal to the boundary. Here we have used the numerical equivalence, first pointed out by Tai [1964a], between the effective quantities $\overline{E}_{eff}$, $\overline{H}_{eff}$ in (6-12) and (6-13), used by Fano, Chu, and Adler [1960], and the starred quantities $\overline{E}^*$, $\overline{H}^*$, used originally by Sommerfeld [1964a] for (6-14). Substituting (6-12) and (6-13) into (6-14), for our present case of no $\hat{z}$ field components, one has for $\hat{n} = \hat{z}$ and $\overline{v}_o = v_o\hat{z}$:

$$[\hat{z} \times \overline{E} - v_o\mu_o\overline{H}]_{z=v_ot-} = [\hat{z} \times \overline{E} - v_o\mu_o\overline{H}]_{z=vt_o+} \tag{6-15}$$

$$[\hat{z} \times \overline{H} + v_o\epsilon_o\overline{E}]_{z=v_ot-} = [\hat{z} \times \overline{H} + v_o\epsilon_o\overline{E}]_{z=vt_o+} \tag{6-16}$$

where $z = v_ot-$ represents the free space side and $z = v_ot+$ represents the plasma side of the drifting boundary.

In order to obey the boundary conditions (6-15) - (6-16) at the drifting boundary at all times, all the waves are required to have the same exponential time variation at $z = v_ot$, as follows:

$$\omega_I t - k_I v_o t = \omega_R t + k_R v_o t = \omega_T t - k_T v_o t \tag{6-17}$$

Using the above definitions, one has the requirement:

$$\omega_I(1-\beta_L) = \omega_R(1+\beta_L) = \omega_T(1-n\beta_L) \qquad (6\text{-}18)$$

Assuming that the source frequency ω_I of the incident wave is given, one finds:

$$\omega_R = \omega_I \frac{1-\beta_L}{1+\beta_L} \; ; \; \omega_T = \omega_I \frac{1-\beta_L}{1-n\beta_L} \qquad (6\text{-}19)$$

Because of the boundary conditions at the drifting boundary, the frequency of the reflected wave ω_R, and the frequency of the transmitted wave in the plasma ω_T, are given in terms of the frequency of the incident wave ω_I by (6-19). Let us define a new set of plasma parameters with respect to the circular frequency ω_I of the incident wave:

$$X_I = \frac{\omega_p^2}{\omega_I^2}, \; Y_I = \frac{\omega_{HL}}{\omega_I}, \; Z_I = \frac{\nu}{\omega_I},$$

and using these together with (6-19) in (6-10), one obtains:

$$(1-\beta_L)(n^2-1)[1-\beta_L - iZ_I \pm Y_I] + X_I(1-n\beta_L)^2 = 0 \qquad (6\text{-}20)$$

Equation (6-20) is a quadratic equation for the refractive index n of the waves in the drifting plasma, as compared to (6-10), which is a cubic equation. In general there will be four different solutions for n in (6-20), two for $-Y_I$ and two for $+Y_I$; one of each pair will represent transmitted characteristic waves in the drifting plasma in the positive $\hat{z}$ direction, and their refractive indices will be denoted by n_1 and n_2, respectively.

Using (6-4) and (6-11), one has for the first transmitted plasma wave ($n = n_1$; $\rho = -i$):

$$\overline{E}_{T1} = E^{(1)}(\hat{x} - i\hat{y})e^{i(\omega_{T1}t - k_{T1}z)} \qquad (6\text{-}21)$$

$$\overline{H}_{T1} = i\frac{n_1}{\eta_o}E^{(1)}(\hat{x} - i\hat{y})e^{i(\omega_{T1}t - k_{T1}z)} \qquad (6\text{-}22)$$

where $E^{(1)}$ is a constant, $\omega_{T1} = \omega_I \frac{1-\beta_L}{1-n_1\beta_L}$ and $k_{T1} = \frac{n_1\omega_{T1}}{c} = \frac{n_1\omega_I}{c}\frac{1-\beta_L}{1-n_1\beta_L}$. For the second transmitted plasma wave ($n = n_2$, $\rho = +i$) one has:

$$\bar{E}_{T2} = E^{(2)}(\hat{x} + i\hat{y})e^{i(\omega_{T2}t - k_{T2}z)} \tag{6-23}$$

$$\bar{H}_{T2} = -i\,\frac{n_2}{\eta_0}\,E^{(2)}(\hat{x} + i\hat{y})e^{i(\omega_{T2}t - k_{T2}z)} \tag{6-24}$$

where $E^{(2)}$ is a constant, $\omega_{T2} = \omega_I\,\dfrac{1 - \beta_L}{1 - n_2\beta_L}$ and $k_{T2} = \dfrac{n_2\omega_{T2}}{c} = \dfrac{n_2\omega_I}{c}\,\dfrac{1 - \beta_L}{1 - n_2\beta_L}$. Both of the waves are circularly polarized, but in opposite directions; because of the boundary conditions at the drifting boundary, the waves will propagate with different circular frequencies ω_{T1}, ω_{T2} in the plasma, for the same frequency ω_I of the incident wave.

Assuming that the amplitude E_x^I of the linearly polarized incident plane wave in (6-1) is known, one can find E_x^R, E_y^R of the reflected waves in (6-2) - (6-3) and $E^{(1)}$, $E^{(2)}$ of the transmitted waves in (6-21) - (6-24), by using the four boundary conditions in (6-15) - (6-16) at the drifting boundary $z = v_0 t$. Substituting (6-1) - (6-3) and (6-21) - (6-24) in the boundary conditions (6-15) - (6-16), cancelling the identical exponential time variations at $z = v_0 t$, taking $v_0\epsilon_0\eta_0 = v_0\mu_0/\eta_0 = v_0/c = \beta_L$, and rearranging, one obtains:

$$\begin{bmatrix} -(1+\beta_L) & 0 & +(1-n_1\beta_L) & +(1-n_2\beta_L) \\ 0 & -i(1+\beta_L) & +(1-n_1\beta_L) & -(1-n_2\beta_L) \\ 0 & +i(1+\beta_L) & +(n_1-\beta_L) & -(n_2-\beta_L) \\ +(1+\beta_L) & 0 & +(n_1-\beta_L) & +(n_2-\beta_L) \end{bmatrix} \begin{bmatrix} E_x^R \\ E_y^R \\ E^{(1)} \\ E^{(2)} \end{bmatrix} = \begin{bmatrix} 1-\beta_L \\ 0 \\ 0 \\ 1-\beta_L \end{bmatrix} E_x^I \tag{6-25}$$

Equation (6-25) can easily be solved for the four unknowns E_x^R, E_y^R, $E^{(1)}$, $E^{(2)}$ in terms of the incident electric field E_x^I; the results will be given in section 6.4. In the next section the relativistic solution will be found, and it will be shown to be identical with equation (6-25). The plasma refractive indices n_1, n_2 in (6-25) are found in the moving bounded magneto-plasma from the double quadratic equation (6-20).

6.3 The Relativistic Solution

In the relativistic method of solution, the incident wave properties in the laboratory system of coordinates are first transformed to the rest system of the moving plasma. In this system of coordinates, the well-known boundary conditions for the tangential electric and magnetic fields are used to find the reflected and the transmitted waves. The wave fields thus found are then transformed to the laboratory system of coordinates, resulting in the same matrix equation (6-25). The solution for the resent case under consideration is found to be identical, by using the non-relativistic method of solution in the previous section 6. , or the relativistic method of solution in the present section 6.3.

Let S denote the laboratory system with coordinate axes x, y, z and let S' denote the moving plasma rest system with coordinate axes x', y', z', parallel to those of S, respectively (Fig. 1). Consider an observer at 0' in S', moving together with the plasma, to whom the plasma would look like a stationary non-moving medium. The Lorentz transformation of coordinates for the present case found from (2-7) and (2-8) is as follows:

$$x' = x; \; y' = y; \; z' = \gamma(z - v_0 t); \; t' = \gamma(t - \frac{v_0}{c^2} z) \qquad (6\text{-}26)$$

In the rest system of coordinates S', embedded in the moving semi-infinite plasma, the circular wave frequency of the reflected wave ω_R' and circular wave frequency of the transmitted wave ω_T' are equal to the circular wave frequency of the incident wave ω_I', observed in this rest system S' by the observer at 0':

$$\omega' = \omega_I' = \omega_R' = \omega_T' \qquad (6\text{-}27)$$

The corresponding circular wave frequencies, observed in the laboratory system of coordinates S, may be found by using the Lorentz transformation (2-26) in (6-27); and then one obtains [Papas, 1965]:

$$\omega' = \gamma\omega_I(1 - \beta_L) = \gamma\omega_R(1 + \beta_L) = \gamma\omega_T(1 - n\beta_L) \qquad (6\text{-}28)$$

where $k_o' = \omega'/c$, $k_o = \omega_I/c$ and $\gamma = (1 - \beta_L^2)^{-1/2}$. The primed quantities will refer to S' and the unprimed quantities to the corresponding terms in S. One finds from (6-28) that in the laboratory system S the circular frequencies of the different waves differ from each other, while they are identical in (6-27) in the rest system of the moving plasma S'. Furthermore, one finds that the relativistic result found in (6-28) is identical with the non-relativistic result found in (6-18) and (6-19) for the present case under consideration.

Let the incident wave fields be E_x^I and H_y^I. Observed in S', the field components will be found from the Lorentz transformations (2-13) and (2-16) as follows [Sommerfeld, 1964a]:

$$E_x'^I = \gamma(E_x^I - v_o\mu_o H_y^I) \tag{6-29}$$

$$H_y'^I = \gamma(H_y^I - v_o\epsilon_o E_x^I) \tag{6-30}$$

where for free space $B_y' = \mu_o H_y'$ and $B_y = \mu_o H_y$. E_x^I and H_y^I are related to each other by the free space wave impedance η_o so that, $E_x^I = \eta_o H_y^I$. Using this relation in (6-29) and (6-30), one obtains:

$$E_x'^I = \gamma(1 - \beta_L)E_x^I \tag{6-31}$$

$$\eta_o H_y'^I = \gamma(1 - \beta_L)E_x^I \tag{6-32}$$

$$E_x'^I = \eta_o H_y'^I \tag{6-33}$$

Equations (6-27) - (6-32) relate the incident wave properties in S' to the incident wave properties in S.

The wave properties in the plasma are determined by the refractive indices. Considering the so-called convection current model [Brandstatter, 1963], the refractive index equation in the rest frame of the plasma is found to be the well-known Appleton-Hartree formula [Budden, 1961]:

$$n'^2 = 1 - \frac{\omega_p'^2}{\omega'^2 - i\nu'\omega' \pm \omega_H'\omega'} \tag{6-34}$$

where $n' = c/u' = \omega'/k_o'u'$ is the refractive index in S'. The plasma parameters in S' in (6-34) are defined as follows:

$\omega_p'^2 = \frac{N'e^2}{\epsilon_0 m'}$; ω_p' = plasma frequency; $e' = e$

ν' = mean collision frequency of the electrons

$\omega_H' = \frac{eB_0}{m'}$ = gyro-frequency of electrons

$B_0 = B_0'$ = magnitude of the static longitudinal magnetic field.

It should be noted from (6-27) and (6-28) that the incident, reflected, and transmitted waves have the same frequency $\omega' = \gamma\omega_I(1 - \beta_L)$ in the S' reference frame. The single fluid continuum theory is assumed in the above formula (6-34) and the plasma waves contain an exponential factor $e^{i(\omega' t' - k_0 n' z')}$. The condition determining (6-34) shows that for the non-resonant case under study $E_z' = 0$. Equation (6-34) gives, in general, one upgoing and one downgoing wave for each sign.

Let n_1' and n_2' be the refractive indices in S' corresponding to the upper and lower signs in (6-34) for the waves in +z' direction. Let n_1 and n_2 be the corresponding refractive indices observed in the laboratory system S, which are related to n_1' and n_2' by the relation [Papas, 1965] found from the Lorentz transformations (2-25) - (2-28):

$$n = \frac{n' + \beta_L}{1 + n'\beta_L} \tag{6-35}$$

which may be rearranged to obtain

$$n' = \frac{n - \beta_L}{1 - n\beta_L} \tag{6-36}$$

One can, therefore, find n as the solutions of a transformed refractive index equation. In order to transform the refractive index equation (6-34) completely into system S, it is necessary to transform the plasma parameters as well. Using the plasma parameters transformations of section 2.4, $\omega_p' = \omega_p$, $\nu' = \gamma\nu$, and $\omega_H' = \gamma\omega_H$, equation (6-34) can be written as:

$$n'^2 = 1 - \frac{\omega_p^2}{\omega'^2 - i\gamma\nu\omega' \pm \gamma\omega_H\omega'} \tag{6-37}$$

where $\omega_p^2 = \frac{Ne^2}{\epsilon_o m}$, and $\omega_H = \frac{eB_o}{m}$. This equation gives n', once the plasma parameters are known in the laboratory frame. By substituting (6-36) in (6-37), one obtains:

$$1-n'^2 = \frac{(1-n^2)(1-\beta_L^2)}{(1-n\beta_L)^2} = \frac{\omega_p^2}{\omega'^2 - i\gamma\nu\omega' \pm \gamma\omega_H\omega'} \tag{6-38}$$

Equation (6-38) explains the difference between the number of the plasma waves that exist in unbounded and bounded drifting magneto-plasmas. In the theory of unbounded, relativistic, drifting magneto-plasma, Unz [1966a] has shown that by taking from (6-28) $\omega' = \gamma\omega_T(1 - n\beta_L)$ and substituting in (6-38), one obtains two cubic equations in n, in terms of the circular frequency of the wave in the plasma ω_T:

$$(n^2 - 1)(1-n\beta_L - i\frac{\nu}{\omega_T} \pm \frac{\omega_H}{\omega_T}) + \frac{\omega_p^2}{\omega_T^2}(1 - n\beta_L) = 0 \tag{6-39}$$

where (6-39) is identical with (6-10) and previous results. However, in the present case of bounded drifting magneto-plasma, because of the boundary conditions, one must take from (6-28) $\omega' = \gamma\omega_I(1 - \beta_L)$, which upon substitution in (6-38), gives two quadratic equations in n, in terms of the circular frequency of the incident wave ω_I:

$$(n^2 - 1)(1 - \beta_L - i\frac{\nu}{\omega_I} \pm \frac{\omega_H}{\omega_I}) + \frac{\omega_p^2}{\omega_I^2}\frac{(1 - n\beta_L)^2}{(1 - \beta_L)} = 0 \tag{6-40}$$

where (6-40) is identical with (6-20). For each sign there will be two waves in the opposite directions.

Let us consider now the incident free space wave in S' represented by :

$$\overline{E}'_I = \hat{x}E'^I_x e^{i(\omega't' - k'_o z')} \tag{6-41}$$

$$\overline{H}'_I = \hat{y}H'^I_y e^{i(\omega't' - k'_o z')} \tag{6-42}$$

and the two reflected free space waves represented by:

$$(\bar{E}'^{(1)}_R,\ \bar{H}'^{(1)}_R) = (\hat{x}E'^R_x,\ \hat{y}H'^R_y)e^{i(\omega't'+k'_oz')} \tag{6-43}$$

$$(\bar{E}'^{(2)}_R,\ \bar{H}'^{(2)}_R) = (\hat{y}E'^R_y,\ \hat{x}H'^R_x)e^{i(\omega't'+k'_oz')} \tag{6-44}$$

and the two transmitted plasma waves represented by:

$$\bar{E}'^{(1)}_T = (\hat{x}E'^{(1)}_x + \hat{y}E'^{(1)}_y)e^{i(\omega't'-k'_on'_1z')} \tag{6-45}$$

$$\bar{H}'^{(1)}_T = (\hat{x}H'^{(1)}_x + \hat{y}H'^{(1)}_y)e^{i(\omega't'-k'_on'_1z')} \tag{6-46}$$

$$\bar{E}'^{(2)}_T = (\hat{x}E'^{(2)}_x + \hat{y}E'^{(2)}_y)e^{i(\omega't'-k'_on'_2z')} \tag{6-47}$$

$$\bar{H}'^{(2)}_T = (\hat{x}H'^{(2)}_x + \hat{y}H'^{(2)}_y)e^{i(\omega't'-k'_on'_2z')} \tag{6-48}$$

where the field components in (6-41) - (6-44) satisfy

$$\frac{E'^I_x}{H'^I_y} = -\frac{E'^R_x}{H'^R_y} = \frac{E'^R_y}{H'^R_x} = \eta_o \tag{6-49}$$

and the field components in (6-45) - (6-48) satisfy [Budden, 1961]:

$$E'^{(1)}_y = -iE'^{(1)}_x;\quad E'^{(2)}_y = iE'^{(2)}_x;\quad i = \sqrt{-1} \tag{6-50}$$

$$\eta_oH'^{(j)}_x = -n'_jE'^{(j)}_y;\quad \eta_oH'^{(j)}_y = n'_jE'^{(j)}_x;\quad j = 1, 2 \tag{6-51}$$

Equation (6-51) is obtained from the Maxwell equation:

$$\nabla' x\bar{E}' = -\mu_o\frac{\partial\bar{H}'}{\partial t'} \tag{6-52}$$

where after introducing the space-time exponential variation of the form $e^{i(\omega't'-k'_on'_jz')}$, one obtains:

$$(-\hat{x}E'^{(j)}_y + \hat{y}E'^{(j)}_x)k'_on'_j = \omega'\mu_o\bar{H}'^{(j)}, \tag{6-53}$$

which in turn yields (6-51). In (6-52) it is assumed that the plasma is non-magnetizable, hence $\bar{B}' = \mu_o\bar{H}'$.

The boundary conditions at z' = 0 may be written as follows for the rest system of the plasma [Landau and Lifshitz, 1966]:

$$E'^{(+0)}_x = E'^{(-0)}_x \tag{6-54}$$

$$E'_y{}^{(+0)} = E'_y{}^{(-0)} \quad (6\text{-}55)$$

$$H'_x{}^{(+0)} = H'_x{}^{(-0)} \quad (6\text{-}56)$$

$$H'_y{}^{(+0)} = H'_y{}^{(-0)} \quad (6\text{-}57)$$

where (+0) indicates a position just above z' = 0, on the plasma side, and (-0) indicates a position just below z' = 0, on the free space side. Substituting (6-41) - (6-48) in (6-54) - (6-57), one obtains:

$$E'_x{}^{(1)} + E'_x{}^{(2)} = E'_x{}^{I} + E'_x{}^{R} \quad (6\text{-}58)$$

$$E'_y{}^{(1)} + E'_y{}^{(2)} = E'_y{}^{R} \quad (6\text{-}59)$$

$$\eta_o H'_x{}^{(1)} + \eta_o H'_x{}^{(2)} = \eta_o H'_x{}^{R} \quad (6\text{-}60)$$

$$\eta_o H'_y{}^{(1)} + \eta_o H'_y{}^{(2)} = \eta_o H'_y{}^{I} + \eta_o H'_y{}^{R} \quad (6\text{-}61)$$

Using the relations (6-49) - (6-50) in (6-58) - (6-61) and rearranging, one obtains the following set of four inhomogeneous linear algebraic equations with four unknowns, in a matrix form, by substituting (6-31) - (6-32):

$$\begin{bmatrix} -1 & 0 & 1 & 1 \\ 0 & -i & 1 & -1 \\ 0 & i & n'_1 & -n'_2 \\ 1 & 0 & n'_1 & n'_2 \end{bmatrix} \begin{bmatrix} E'_x{}^{R} \\ E'_y{}^{R} \\ E'_x{}^{(1)} \\ E'_x{}^{(2)} \end{bmatrix} = \begin{bmatrix} (1-\beta_L) \\ 0 \\ 0 \\ (1-\beta_L) \end{bmatrix} \gamma E_x^{I} \quad (6\text{-}62)$$

Equation (6-62) can be solved for reflected and transmitted fields in S' in terms of the incident electric field E_x^I, once n'_1 and n'_2 are known. The reflected and transmitted fields in S' can, however, be written in terms of the reflected and transmitted fields in S as follows [Sommerfeld, 1964a], by using the Lorentz transformations (2-13) and (2-16):

$$E'_x{}^{R} = \gamma(1+\beta_L)E_x^{R} \quad (6\text{-}63)$$

$$E'_y{}^{R} = \gamma(1+\beta_L)E_y^{R} \quad (6\text{-}64)$$

$$E_x'^{(1)} = \gamma(1 - n_1\beta_L)E^{(1)} \qquad (6\text{-}65)$$

$$E_x'^{(2)} = \gamma(1 - n_2\beta_L)E^{(2)} \qquad (6\text{-}66)$$

where the unit vectors $\hat{x}$, $\hat{y}$, $\hat{z}$ are identical with the unit vectors $\hat{x}'$, $\hat{y}'$, $\hat{z}'$, respectively, and assuming that representations similar to (6-45) - (6-48) and results similar to (6-51) exist in the S frame, derivable from $\nabla x\bar{E} = -i\omega_j\mu_0\bar{H}$. It should be noted that in the convection current model used in this section, one has $\bar{D} = \epsilon_0\bar{E}$ and $\bar{B} = \mu_0\bar{H}$ in the plasma. This is a result of considering $\epsilon' = \epsilon_0$ and $\mu' = \mu_0$ in the rest system of the plasma. Substituting (6-36) and (6-63) - (6-66) in (6-62) and cancelling the common factor γ, one obtains equation (6-25). Thus equation (6-25) is true as well for relativistic velocities. Analytic expressions for the reflected and transmitted waves in terms of the incident wave are given in the following section.

6.4 Reflection and Transmission Coefficients

Equation (6-25) can be solved by inverting the matrix, and the reflection and transmission coefficients defined as follows are found to be:

$$\frac{E_x^R}{E_x^I} = r_1 = \frac{1}{2}\left[\frac{1-n_1}{1+n_1} + \frac{1-n_2}{1+n_2}\right] = \frac{1-n_1n_2}{(1+n_1)(1+n_2)} \qquad (6\text{-}67)$$

$$\frac{E_y^R}{E_x^I} = r_2 = i\left[\frac{1}{1+n_2} - \frac{1}{1+n_1}\right] = \frac{i(n_1 - n_2)}{(1+n_1)(1+n_2)} \qquad (6\text{-}68)$$

$$\frac{E^{(1)}}{E_x^I} = t^{(1)} = \frac{1}{(1+n_1)} \qquad (6\text{-}69)$$

$$\frac{E^{(2)}}{E_x^I} = t^{(2)} = \frac{1}{(1+n_2)} \qquad (6\text{-}70)$$

where (6-67) - (6-70) apply to both the non-relativistic [Unz, 1967a] and the relativistic [Chawla and Unz, 1967a] moving plasmas ($\beta_L \neq 0$), and are identical with the results given by Budden [1961] for the case of non-drifting magneto-plasma ($\beta_L = 0$). Equations (6-67) - (6-70) give the electric fields of the reflected and transmitted waves in terms of the incident wave electric field and depend on $\beta_L \neq 0$ only through (6-20) or (6-40).

The power reflection coefficient for the semi-infinite magneto-plasma may be defined by:

$$R = -\hat{n}\cdot\overline{S}_r/\hat{n}\cdot\overline{S}_i \qquad (6\text{-}71)$$

where $\hat{n}$ is the unit vector normal to the surface and $\overline{S}_r$, $\overline{S}_i$ are the corresponding Poynting vectors in free space defined by:

$$\overline{S}_r = \frac{1}{2}(\overline{E}_R^{(1)} \times \overline{H}_R^{(1)*} + \overline{E}_R^{(2)} \times \overline{H}_R^{(2)*}) \qquad (6\text{-}72)$$

$$\overline{S}_i = \frac{1}{2}(\overline{E}_I \times \overline{H}_I^*) \qquad (6\text{-}73)$$

where * signifies the complex conjugate of the function. The transmission coefficient cannot be defined as in (6-71), since in general the transmitted wave frequency will be complex [Yeh, 1966]. The above definition is correct for real frequencies only.

Using the results of the previous section, one may obtain for $\hat{n} = \hat{z}$:

$$\hat{n}\cdot\overline{S}_r = \frac{-1}{2\eta_o}\left[|E_x^R|^2 + |E_y^R|^2\right] \qquad (6\text{-}74)$$

$$\hat{n}\cdot\overline{S}_i = \frac{1}{2\eta_o}|E_x^I|^2 \qquad (6\text{-}75)$$

Substituting (6-74) - (6-75) in (6-71) and (6-67) -(6-68) in the resulting expression, one obtains:

$$R = |r_1|^2 + |r_2|^2 \qquad (6\text{-}76)$$

The power reflection coefficient R will be zero when $\beta_L \to 1.0$, and the plasma is moving away from the observer at the velocity of light. However, when $\beta_L \to -1.0$, and the plasma is moving towards the observer at the velocity of light, the reflection coefficients r_1 and r_2 in (6-67) - (6-80) reduce asymptotically, by using (6-40) and $\nu = 0$, to:

$$r_1 = \frac{\omega_p^2}{(4\omega_I^2 - \omega_H^2)}\;;\; r_2 = -\frac{i}{2}\,\frac{\omega_H\omega_p^2}{\omega_I(4\omega_I^2 - \omega_H^2)}, \qquad (6\text{-}77)$$

and the power reflection coefficient R will then reduce asymptotically to:

$$R = \frac{(4\Omega^2 + \Omega_H^2)}{4(4\Omega^2 - \Omega_H^2)^2\,\Omega^2} \tag{6-78}$$

where $\Omega = \omega_I/\omega_p$ and $\Omega_H = \omega_H/\omega_p$.

For the particular case of no magnetic field ($\Omega_H = 0$), equation (6-78) reduces to :

$$R = (1/16)(\omega_p/\omega_I)^4 \tag{6-79}$$

One will note that in the paper by Yeh [1966] the factor (1/16) in (6-79) is missing and should be added [Yeh, Private Communication].

6.5 Numerical Results and Discussion

Figures 2, 3, and 4 are the plots of the reflection coefficient R vs. the velocity of the medium β_L, with $\Omega = \omega_I/\omega_p = \omega/\omega_p$ as a parameter, where $\omega \equiv \omega_I$. Computations were performed for intervals of $\beta_L = 0.005$. Figures 5, 6, and 7 are the plots of R vs. the normalized wave frequency $\Omega = \omega_I/\omega_p$, with β_L as a parameter. These computations were performed for intervals of $\Omega = 0.02$. The effect of collisions was neglected by taking $\nu = 0$. The different graphs are for different values of the magnetic field, the parameter being the normalized gyrofrequency $\Omega_H = \omega_H/\omega_p$.

The computations involved finding n' from the equations (6-28) and (6-37):

$$n'^2 = 1 - \frac{(1 + \beta_L)}{(1 - \beta_L)\Omega^2 \pm \Omega\Omega_H} \tag{6-80}$$

where the sign of n' was chosen so that the waves in S' propagate or attenuate in the +z' direction; then applying the transformation (6-35):

$$n = (n' + \beta_L)/(1 + n'\beta_L) \tag{6-81}$$

to obtain n, and finally substituting n in the expression for the reflection coefficient (6-76), using (6-67) and (6-68).

Figure 2 is for $\Omega_H = 0$, where the external magnetic field is zero. Various curves are given for different values of $(\omega_p/\omega_I)^2$. The different values of $(\omega_p/\omega_I)^2$ were taken to be the same as in the paper by Yeh [1966] for a comparison. The comparison shows a disagreement in the value of the maxima, the values of β_L at which maxima occur, when taking into account the reversal in the notation for β_L, and various other features. Theoretically, the present results for the anisotropic moving plasma reduce to his formulas for the particular case of no magnetic field. The numerical results show that the portion of one curve is the same as that of the other for all the cases of ω_p/ω_I, instead of only for $\omega_p/\omega_I > 1$ as indicated by Yeh [1966]. The other feature indicated by Yeh, "The reflection coefficient no longer increases monotonically as β_z increases for all values of ω_p/ω_I," is in agreement. However, it should be noted that the reflection coefficient for $\omega_p/\omega_I > 1$ is larger than unity only in a certain range of β_L, rather than from $\beta_L = 0$ to $\beta_L = -1$. This would be so for $\omega_p/\omega_I > 4$ only, since for $\omega_p/\omega_I < 4$ the asymptotic value is less than unity.

Figures 3 and 4 show the effect of the magnetic field on the reflection coefficient. Figure 3 is for $\Omega_H = 0.5$, and the characteristics are essentially the same as in figure 2; however, it is interesting to note that for each ω_p/ω_I the maxima are higher than the corresponding maxima for the case of no magnetic field. Each of the curves here have one more maximum, which is considerably smaller than the other one. The asymptotic values for each case are higher than the ones for the case of no magnetic field.

Figure 4 is for $\Omega_H = 2.0$. The plots for reflection coefficient are much different from those of figure 3. For $\omega_p/\omega_I > 1$ the reflection coefficient increases monotonically as β_L varies from 0 to -1.0. For $(\omega_p/\omega_I)^2 = 1.5$ the asymptotic maximum is greater than unity. The results for $\omega_p/\omega_I < 1$ are more interesting. For $(\omega_p/\omega_I)^2 = 0.95$ the maximum occurs at $\beta_L = -0.97$, and the value of R at this point is 2.16×10^3. For $(\omega_p/\omega_I)^2 = 0.5$, the maximum occurs at $\beta_L = -0.605$ and has the value 8.26. For experimental purposes the region between these two values of $(\omega_p/\omega_I)^2$ may be of great interest. In figures 2, 3, and 4, $\omega \equiv \omega_I$.

Figure 5 represents the plot of R vs. Ω for $\Omega_H = 0$. The various curves are for different values of

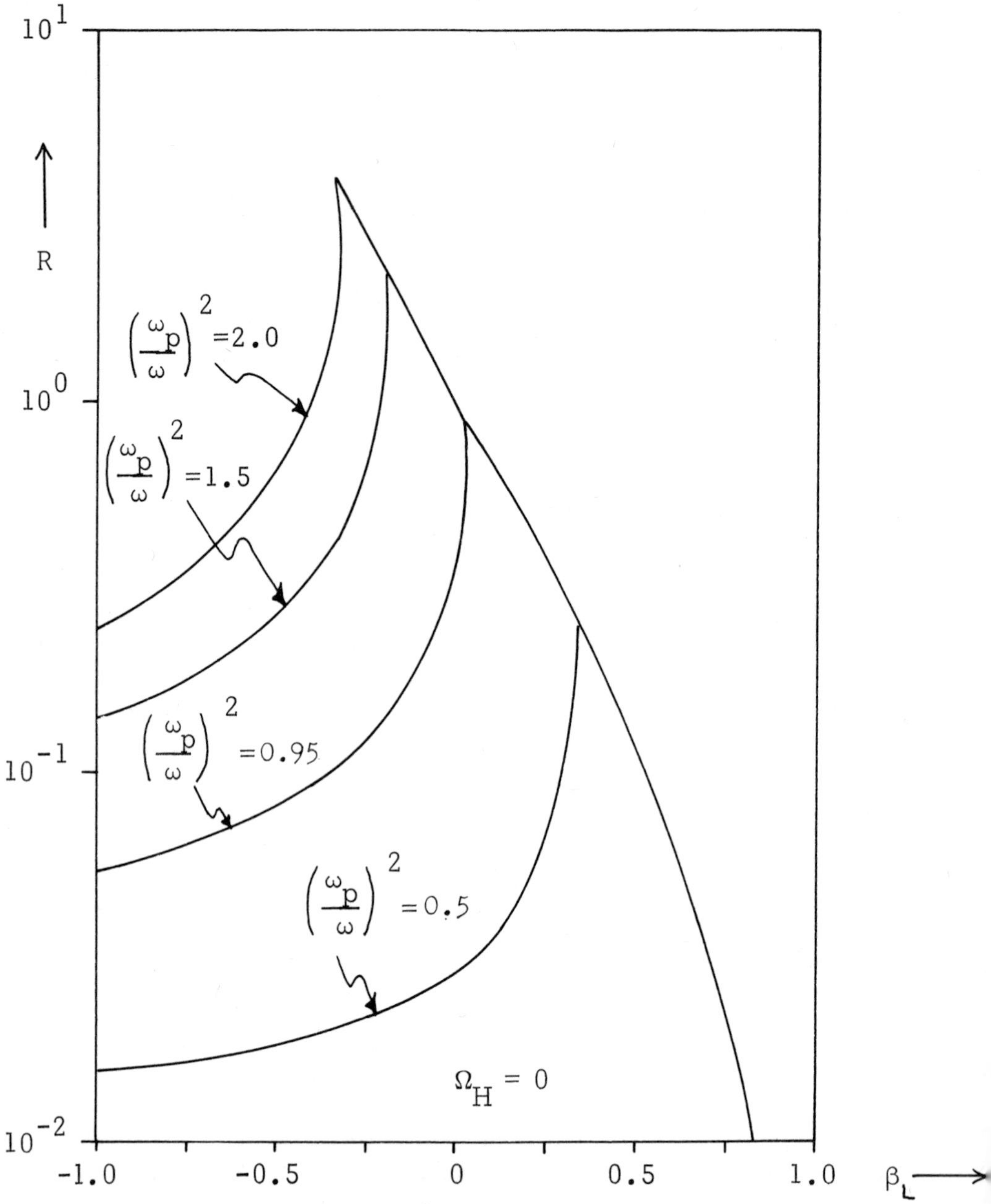

Figure 2. Reflection Coefficient vs. Normalized Velocity for $\Omega_H = 0$

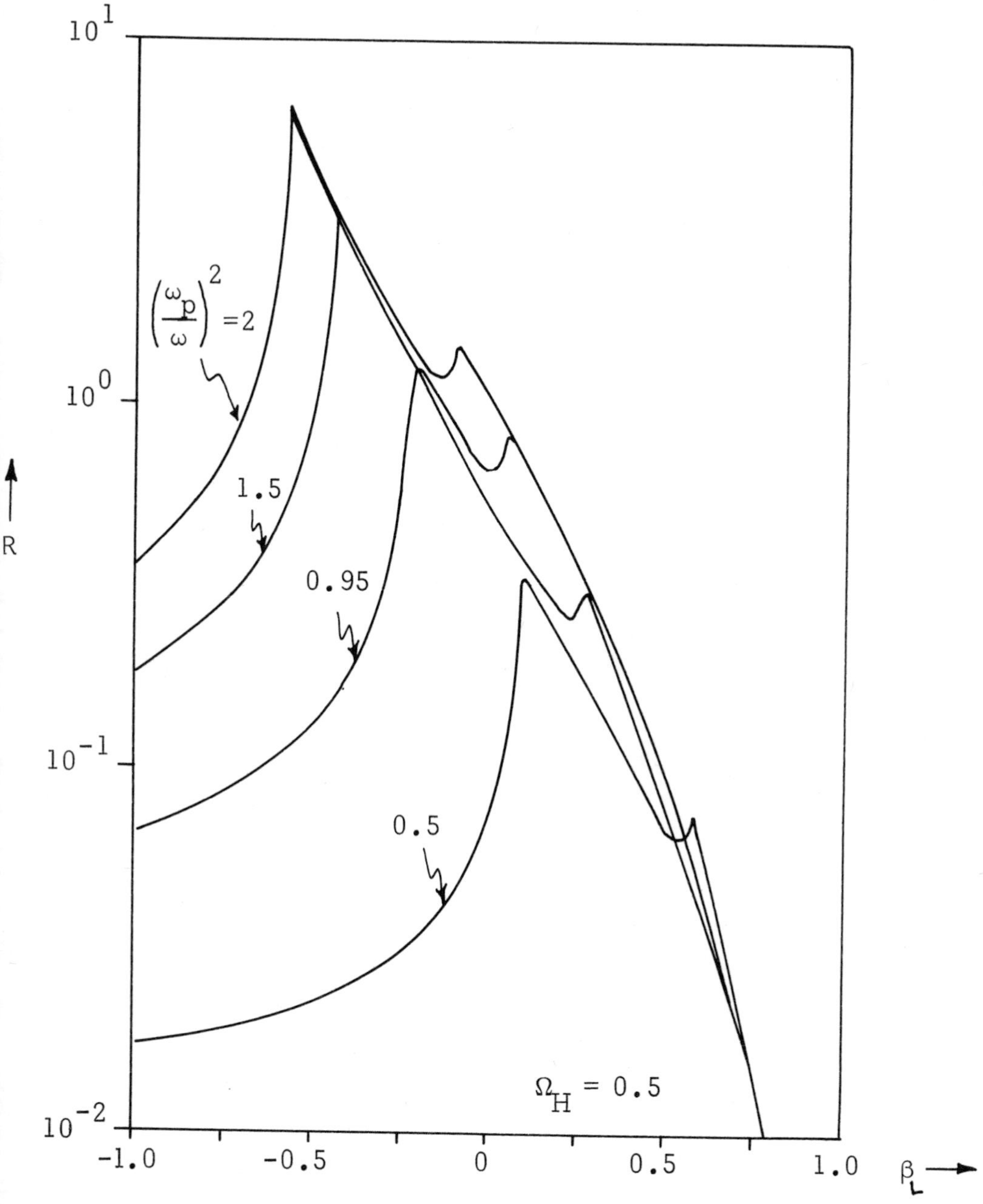

Figure 3. Reflection Coefficient vs. Normalized Velocity for $\Omega_H = 0.5$

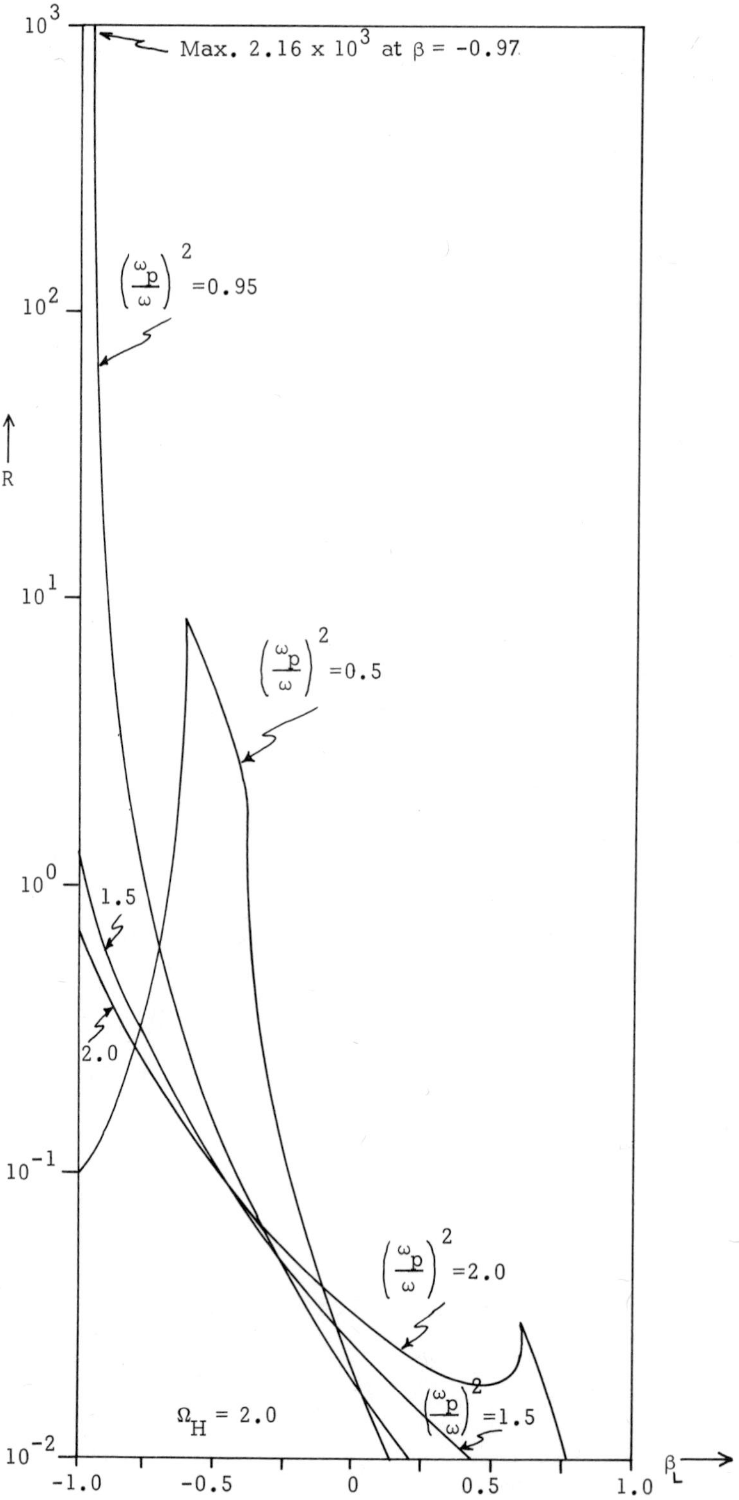

Figure 4. Reflection Coefficient vs. Normalized Velocity for Ω_H = 2.0

β_L. For $\beta_L = 0$ it is noted that the reflection coefficient is unity for $\Omega < 1$; that is, the wave is completely reflected. This result is well known, since in this case $n' = n$ is purely imaginary, and thus represents an evanescent wave in the plasma. For $\beta_L \neq 0$ it is seen that the reflection coefficient takes different constant values, greater than unity for negative values of β_L, and smaller than unity for positive values of β_L. The region with constant R depends on the value of β_L, this constant value is equal to $(1 - \beta_L)^2/(1+\beta_L)^2$. Let Ω_{cr} be the value of Ω up to which R is constant, then Ω_{cr} decreases as $\beta_L \to -1$. It reaches zero at $\beta_L = -1$, where the reflection coefficient is infinite. The result that the reflection coefficient for the negative values of β_L is greater than unity may appear to be surprising at first, since the wave refractive index n' is purely imaginary in the S' system. However, the transformation to n gives a complex value, hence the wave appears to be propagating and attenuating in the S syteem. This is physically reasonable, since due to the motion of the plasma together with its interface, the time maximum of the attenuating wave in S' will appear as propagation. The reflection coefficient being greater than unity can be understood in terms of the transfer of energy from the motion of the plasma to the electromagnetic wave.

Figures 6 and 7 show the effect of the magnetic field on R for different values of β_L. In figure 6 we have $\Omega_H = 0.5$. Here for $\beta_L = 0$ the reflection coefficient is unity between $\Omega = 0.5$ and $\Omega = 0.78$, since both waves in the plasma are evanescent in this region [Seshadri, 1964]. The power reflection coefficient R is different from unity for the rest of values of Ω since one of the waves (extraordinary mode) propagates for $\Omega < 0.5$. The other wave (ordinary mode) propagates for $0.78 < \Omega < 1.28$, while both waves propagate for $\Omega > 1.28$. The rest of the curves for $\beta_L \neq 0$ show the effect of motion of the semi-infinite plasma on the reflection coefficient R. For $\beta_L = -1.0$ it is seen that $R \to \infty$ at $\Omega = 0.25$. This can be seen from the asymptotic expression for R in (6-78) where the denominator is zero when $\Omega = \Omega_H/2$. Also we have $R \to \infty$ as $\Omega \to 0$.

In figure 7 we have $\Omega_H = 2.0$, the rest of the parameters being the same as in figures 5 and 6. The curve for $\beta_L = 0$ is different from that for $\Omega_H = 0.5$, in that now the ordinary mode propagates for $\Omega > 0.414$,

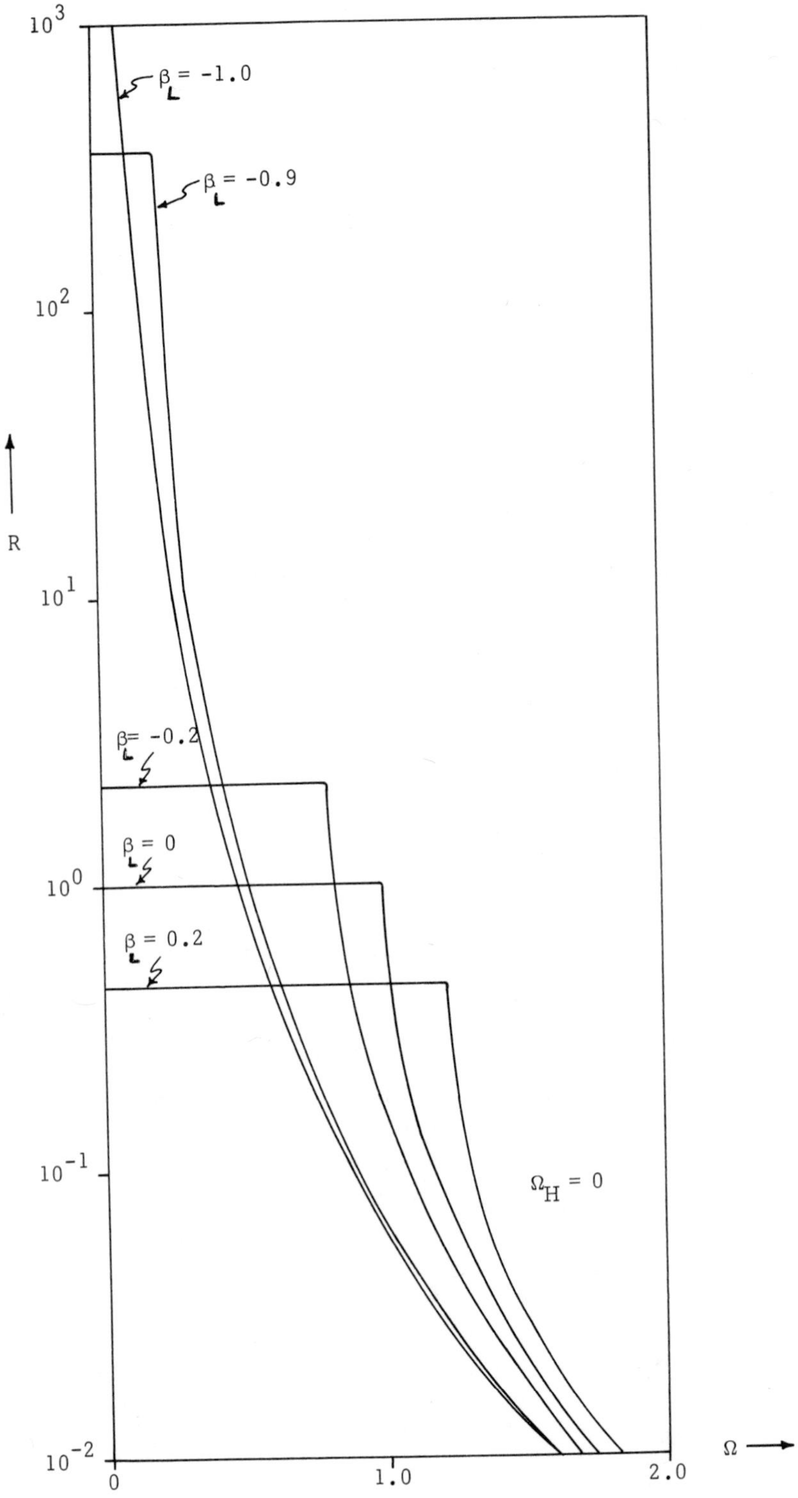

Figure 5. Reflection Coefficient vs. Normalized Incident Wave Frequency for $\Omega_H = 0$

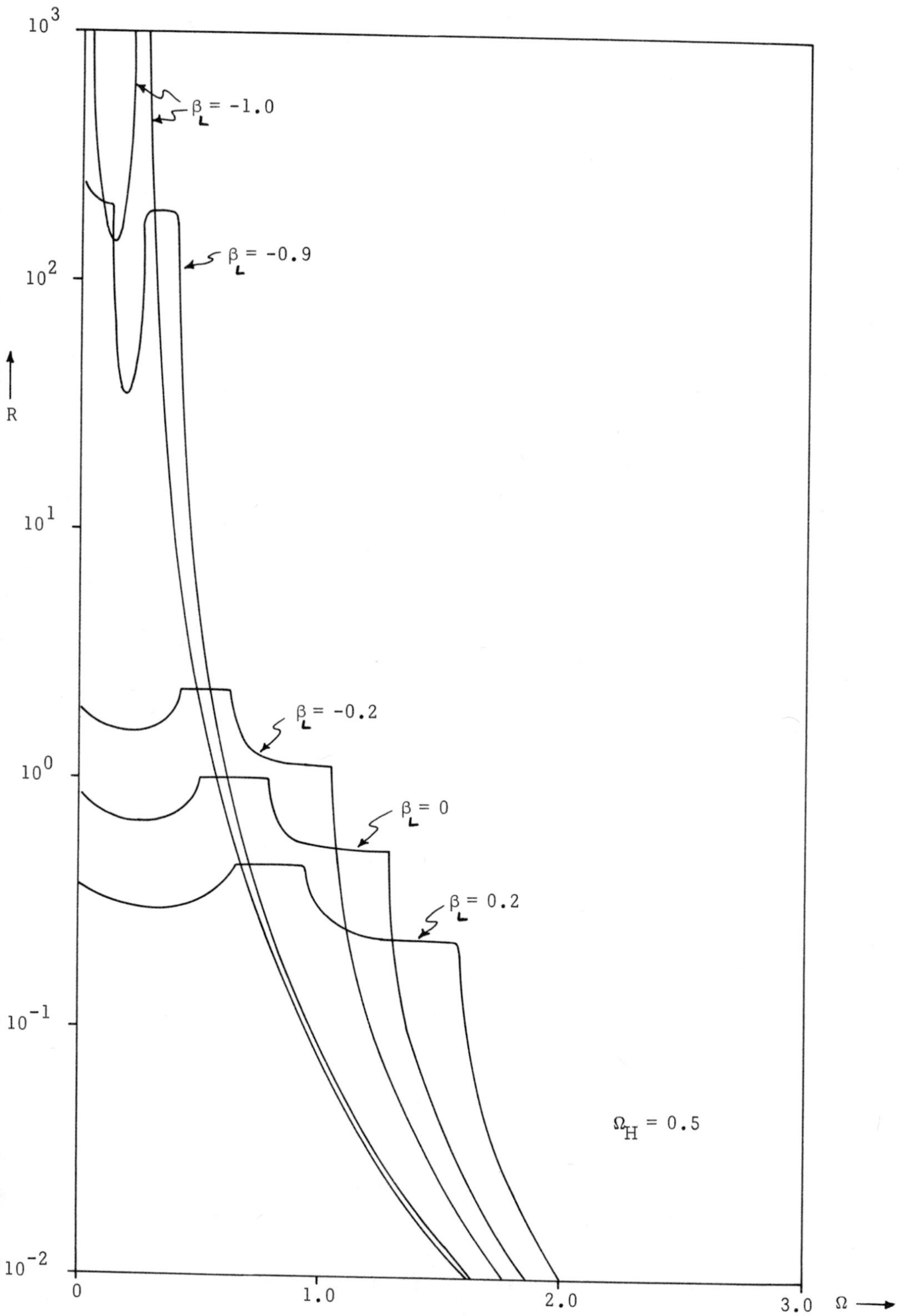

Figure 6. Reflection Coefficient vs. Normalized Incident Wave Frequency for $\Omega_H = 0.5$

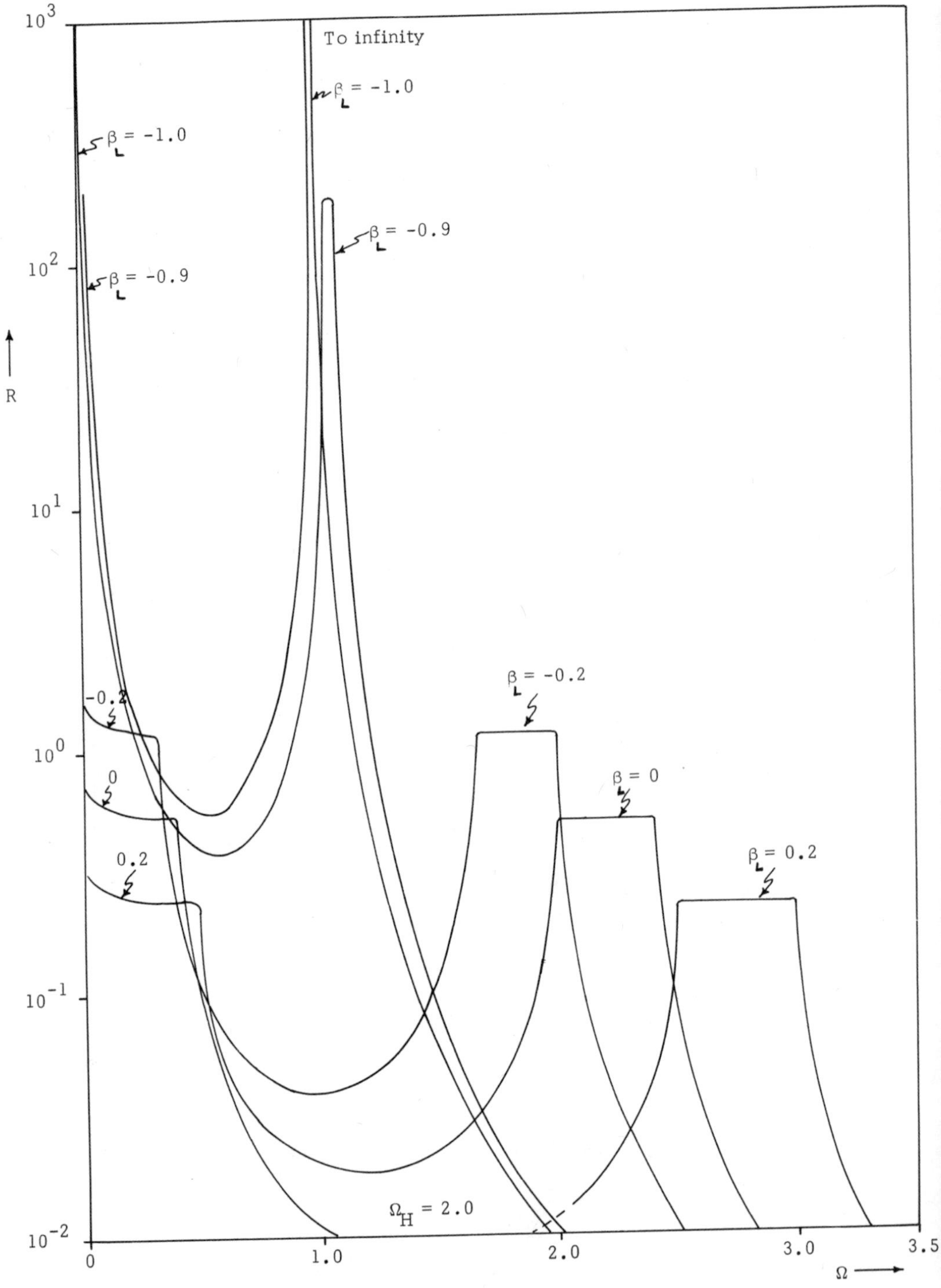

Figure 7. Reflection Coefficient vs. Normalized Incident Wave Frequency for $\Omega_H = 2.0$

while the extraordinary mode propagates for $2.414 < \Omega < 2.0$. There is no region in which both of the waves do not propagate. The reflection coefficient 0.5 for $2.0 < \Omega < 2.414$ is thus a result of the propagation of only the ordinary wave. Modifications due to the motion are shown in the other curves. The region of constant value of R decreases for negative values of β_L as before. For $\beta_L = -1.0$, $R \to \infty$ at $\Omega = 0., 1.0$. For $\beta_L = -0.9$ the peak is narrower and steeper compared to the one for $\Omega_H = 0.5$.

6.6 Circularly Polarized Waves

In the present section we will discuss the reflection and transmission coefficients for a circularly polarized electromagnetic wave, normally incident on a semi-infinite plasma, moving along a uniformly static magnetic field. Since a circularly polarized wave excites a transmitted wave of the same polarization, the polarization of the reflected wave will also be circular. The final result in this case in a closed form, similar to the result for the isotropic case.

Let two linearly polarized incident electromagnetic waves in free space be given by:

$$\bar{E}_{I1} = E_x^I \hat{x} e^{i(\omega_I t - k_I z)} \quad ; \quad \bar{H}_{I1} = \frac{1}{\eta_0} E_x^I \hat{y} e^{i(\omega_I t - k_I z)} \tag{6-82}$$

$$\bar{E}_{I2} = E_y^I \hat{y} e^{i(\omega_I t - k_I z)} \quad ; \quad \bar{H}_{I2} = -\frac{1}{\eta_0} E_y^I \hat{x} e^{i(\omega_I t - k_I z)} \tag{6-83}$$

where E_x^I and E_y^I are constants and $k_I = \omega_I / c$. Equation (6-82), like (6-1), represents a linearly polarized incident wave in the $\hat{x}$ direction, and (6-83) represents a linearly polarized incident wave in the $\hat{y}$ direction.

The drifting gyrotropic plasma boundary, assumed to be located at $z = v_0 t$, will produce in general two eletromagnetic linearly polarized reflected waves in free space, given by (6-2) and (6-3), and two transmitted circularly polarized electromagnetic waves in the plasma, given by (6-21) - (6-24).

In order to obey the boundary conditions (6-15) - (6-16) at the moving boundary $z = v_0 t$ of the semi-infinite plasma, all the waves will be required to have the same exponential time variation as in (6-17)-(6-19), with the resulting plasma quadratic refractive index equation in (6-20). Assuming that the amplitudes E_x^I and E_y^I of the linearly polarized incident plane electromagnetic waves in free space in (6-82) and (6-83) are known, one can find E_x^R and E_y^R of the reflected waves in (6-2) - (6-3) and $E^{(1)}$ and $E^{(2)}$ of the plasma transmitted waves in (6-21) - (6-24), by using the four boundary conditions in (6-15) - (6-16) at the moving plasma boundary $z = v_0 t$.

Substituting (6-2) - (6-3), (6-21) - (6-24), and (6-82) - (6-83) in the boundary conditions (6-15) - (6-16), cancelling the identical exponential time variation at $z = v_0 t$, taking $v_0 \epsilon_0 \eta_0 = v_0 \mu_0 / \eta_0 = v_0/c = \beta_L$, and rearranging, one obtains:

$$-(1+\beta_L)E_x^R + (1-n_1\beta_L)E^{(1)} + (1-n_2\beta_L)E^{(2)} = (1-\beta_L)E_x^I \quad (6\text{-}84)$$

$$-(1+\beta_L)iE_y^R + (1-n_1\beta_L)E^{(1)} - (1-n_2\beta_L)E^{(2)} = (1-\beta_L)iE_y^I \quad (6\text{-}85)$$

$$+(1+\beta_L)iE_y^R + (n_1-\beta_L)E^{(1)} - (n_2-\beta_L)E^{(2)} = (1-\beta_L)iE_y^I \quad (6\text{-}86)$$

$$+(1+\beta_L)E_x^R + (n_1-\beta_L)E^{(1)} + (n_2-\beta_L)E^{(2)} = (1-\beta_L)E_x^I \quad (6\text{-}87)$$

Taking the particular case $E_y^I = 0$ in (6-84) - (6-87), one obtains (6-25), which has been discussed in the previous sections. The plasma refractive indices n_1, n_2 above are found from the double quadratic equation (6-20) for the wave transmitted in the positive $\hat{z}$ direction in the moving bounded magneto-plasma. The four equations (6-84) - (6-87) can be easily solved for the four unknowns E_x^R, E_y^R, $E^{(1)}$, and $E^{(2)}$ in terms of the electric fields of the incident waves E_x^I, and E_y^I. The same equations (6-84) - (6-87) apply also for the relativistic case in accordance with the discussion in section 6.3.

Taking the sum and the difference of (6-84) and (6-85) one has:

$$-(1+\beta_L)(1/2)(E_x^R + iE_y^R) + (1-n_1\beta_L)E^{(1)} = (1-\beta_L)(1/2)(E_x^I + iE_y^I) \quad (6\text{-}88)$$

$$-(1+\beta_L)(1/2)(E_x^R - iE_y^R) + (1-n_2\beta_L)E^{(2)} = (1-\beta_L)(1/2)(E_x^I - iE_y^I) \quad (6\text{-}89)$$

Taking the sum and the difference of (6-87) and (6-86) one has:

$$+(1+\beta_L)(1/2)(E_x^R+iE_y^R)+(n_1-\beta_L)E^{(1)}=(1-\beta_L)(1/2)(E_x^I+iE_y^I) \tag{6-90}$$

$$+(1+\beta_L)(1/2)(E_x^R-iE_y^R)+(n_2-\beta_L)E^{(2)}=(1-\beta_L)(1/2)(E_x^I-iE_y^I) \tag{6-91}$$

Let us assume that the normally incident plane electromagnetic wave is a right-handed circularly polarized wave in free space as follows:

$$\overline{E}_{Ir} = E_r^I\,(\hat{x}-i\hat{y})e^{i(\omega_I t - k_I z)} \tag{6-92}$$

$$\overline{H}_{Ir} = \frac{i}{\eta_o}\,E_r^I\,(\hat{x}-i\hat{y})e^{i(\omega_I t - k_I z)} \tag{6-93}$$

where E_r^I = constant, $k_I = \omega_I/c$, $\eta_o = \sqrt{\mu_o/\epsilon_o}$, and $\rho = E_y/E_x = -i$ represents a clockwise rotation of the fields when one looks in the direction of propagation of the electromagnetic wave $\hat{z}$, and thus is defined [Budden, 1961] as a right-hand circularly polarized wave.

The left-handed circularly polarized normally incident plane electromagnetic wave is given by:

$$\overline{E}_{I\ell} = E_\ell^I\,(\hat{x}+i\hat{y})e^{i(\omega_I t - k_I z)} \tag{6-94}$$

$$\overline{H}_{I\ell} = -\frac{i}{\eta_o}\,E_\ell^I\,(\hat{x}+i\hat{y})e^{i(\omega_I t - k_I z)} \tag{6-95}$$

where E_ℓ^I = constant, and $\rho = E_y/E_x = +i$ represents a counterclockwise rotation of the fields when one looks in the direction of propagation of the electromagnetic wave $\hat{z}$, and thus is defined [Budden, 1961] as a left-handed circularly polarized wave.

The corresponding right-handed circularly polarized electromagnetic wave in the longitudinally moving magneto-plasma is given in (6-21) - (6-22), and the corresponding left-handed circularly polarized electromagnetic wave in the longitudinally moving magneto-plasma is given in (6-23) - (6-24), similarly to the incident circularly polarized waves in free space.

The normally reflected plane electromagnetic right-handed circularly polarized wave in free space is given by :

$$\overline{E}_{Rr} = E_r^R(\hat{x}+i\hat{y})\,e^{i(\omega_R t + k_R z)} \tag{6-96}$$

$$\overline{H}_{Rr} = \frac{i}{\eta_o}\,E_r^R(\hat{x}+i\hat{y})\,e^{i(\omega_R t + k_R z)} \tag{6-97}$$

where E_r^R = constant, $k_R = \omega_R/c$. Equations (6-96) - (6-97) represent a counterclockwise rotation of the fields when one looks in the positive $\hat{z}$ direction, and thus represent a clockwise rotation of the fields when one looks in the direction of propagation of the reflected electromagnetic wave in the negative $\hat{z}$ direction; therefore this wave is defined [Budden, 1961] as a right-handed circularly polarized wave. Similarly, the normally reflected left-handed circularly polarized wave in free space is given by :

$$\overline{E}_{R\ell} = E_\ell^R(\hat{x} - i\hat{y})e^{i(\omega_R t + k_R z)} \tag{6-98}$$

$$\overline{H}_{R\ell} = -\frac{i}{\eta_o} E_\ell^R(\hat{x} - i\hat{y})\, e^{i(\omega_R t + k_R z)} \tag{6-99}$$

where E_ℓ^R = constant. Equations (6-98) - (6-99) represent a clockwise rotation of the field when one looks in the positive $\hat{z}$ direction, but represent a counterclockwise rotation of the fields when one looks in the direction of propagation of the reflected electromagnetic wave in the negative $\hat{z}$ direction; therefore this wave is defined [Budden, 1961] as a left-handed circularly polarized wave. From (6-92) and (6-94) one has for circularly polarized incident waves:

$$(1/2)(E_x^I + iE_y^I) = (1/2)[(E_r^I + E_\ell^I) + i(-iE_r^I + iE_\ell^I)] = E_r^I \tag{6-100}$$

$$(1/2)(E_x^I - iE_y^I) = (1/2)[(E_r^I + E_\ell^I) - i(-iE_r^I + iE_\ell^I)] = E_\ell^I \tag{6-101}$$

From (6-96) and (6-98) one has for circularly polarized reflected waves:

$$(1/2)(E_x^R + iE_y^R) = (1/2)[(E_r^R + E_\ell^R) + i(iE_r^R - iE_\ell^R)] = E_\ell^R \tag{6-102}$$

$$(1/2)(E_x^R - iE_y^R) = (1/2)[(E_r^R + E_\ell^R) - i(iE_r^R - iE_\ell^R)] = E_r^R \tag{6-103}$$

Substituting (6-100) and (6-102) in (6-88) and (6-90), one obtains the following pair of equations:

$$-(1 + \beta_L)E_\ell^R + (1 - n_1\beta_L)E^{(1)} = (1 - \beta_L)E_r^I \tag{6-104}$$

$$(1 + \beta_L)E_\ell^R + (n_1 - \beta_L)E^{(1)} = (1 - \beta_L)E_r^I \tag{6-105}$$

Substituting (6-101) and (6-103) in (6-89) and (6-91), one obtains the following pair of equations:

$$-(1+\beta_L)E_r^R + (1-n_2\beta_L)E^{(2)} = (1-\beta_L)E_\ell^I \qquad (6\text{-}106)$$

$$(1+\beta_L)E_r^R + (n_2-\beta_L)E^{(2)} = (1-\beta_L)E_\ell^I \qquad (6\text{-}107)$$

The pair of equations (6-104) - (6-105) includes only circularly polarized electromagnetic waves in the same sense, i.e. right-handed in the positive $\hat{z}$ direction $E_r^I, E^{(1)}$ and left-handed in the negative $\hat{z}$ direction E_ℓ^R. The pair of equations (6-106) - (6-107) includes only circularly polarized electromagnetic waves in the same sense, i.e. left-handed in the positive $\hat{z}$ direction $E_\ell^I, E^{(2)}$ and right-handed in the negative $\hat{z}$ direction E_r^R. A circularly polarized, normally incident electromagnetic wave on the moving semi-infinite magneto-plasma in the longitudinal direction along the magneto-static field, will cause reflected circularly polarized wave and plasma circularly polarized wave with the same sense of rotation.

Solving the pair (6-104) - (6-105), one obtains:

$$\frac{E_\ell^R}{E_r^I} = r_{r\ell} = \frac{1-n_1}{1+n_1} \; ; \; \frac{E^{(1)}}{E_r^I} = t_{rr} = \frac{2}{1+n_1} \qquad (6\text{-}108)$$

Solving the pair (6-106) - (6-107), one obtains:

$$\frac{E_r^R}{E_\ell^I} = r_{\ell r} = \frac{1-n_2}{1+n_2} \; ; \; \frac{E^{(2)}}{E_\ell^I} = t_{\ell\ell} = \frac{2}{1+n_2} \qquad (6\text{-}109)$$

where $r_{rr} = r_{\ell\ell} = t_{r\ell} = t_{\ell r} = 0$ in accordance with the discussion above, since a I the waves should be circularly polarized in the same sense. Equations (6-108) - (6-109) are identical to the results given by Budden [1961] for the stationary plasma, except for a negative sign, which is the result of the different sign conventions used for the reflected wave electric field.

By comparing (6-108) - (6-109) with (6-67) - (6-68), one may see that by taking a circularly polarized incident electromagnetic wave, the moving gyrotropic plasma medium for the longitudinal case behaves as if it were only an isotropic plasma with only one refractive index n_1 or n_2. Equations (6-108) - (6-109) give the circularly polarized electric fields of the reflected and transmitted waves in terms of the circularly polarized electric fields of the incident waves and depend on $\beta_L \neq 0$ only through (6-20) or (6-40).

While in the present section 6.6, we have discussed the circularly polarized wave solutions based on the non-relativistic solution given in section 6.2, one may obtain identical results by basing the present analysis on the relativistic solution given in section 6.3. The results in the present section 6.6, for the circularly polarized waves, apply for the relativistic as well as for the non-relativistic drift velocities of the magneto-plasma, for the present longitudinal problem under consideration.

VII

NORMAL INCIDENCE ON SEMI-INFINITE TEMPERATE MAGNETO-PLASMA MOVING THROUGH A DIELECTRIC

7.1 Introduction

In the preceding chapter 6, the reflection and the transmission of linearly polarized electromagnetic waves, normally incident on a semi-infinite magneto-plasma, moving through free space, was considered. The results showed that significant effects will be observed when the velocity of the moving plasma is comparable to the velocity of light in free space. The restriction which requires relativistic velocities, however, could be removed, if the plasma were to be moving through a dielectric, since the Doppler effect on frequency depends on the relationship between the velocity of the reflecting body and the velocity of light in the stationary medium. However, large losses would occur if the plasma beam were actually to move through a natural dielectric; therefore, in an experiment, a retarding system, such as a slow-wave waveguide structure, may be used and represented by an equivalent dielectric constant for the sake of simplicity in the analysis.

The use of the non-relativistic beams in slow-wave structures, in the absence of external magnetic field, was considered by Lampert [1956]. In the interaction region, the phase velocity of the electromagnetic wave is reduced and is comparable with the beam velocity. Significant changes in amplitude and frequency may be attained in this case. For analytical simplicity, the mathematical model considered by Lampert [1956] was a semi-infinite plasma moving through a dielectric, on which linearly polarized electromagnetic waves are normally incident, where the dielectric is assumed to be non-dispersive in the frequency range of interest.

An extension of the above, in order to include the effect of the presence of a static magnetic field, was presented by Fainberg and Tkalich [1959]. The direction of the external magnetic field, the velocity of the plasma, and the direction of the incident electromagnetic wave were chosen to be all in the same longitudinal direction, normal to the moving plasma boundary. The incident electromagnetic wave was assumed to be circularly polarized, as was done by Landecker [1952] for the case of plasma moving through free space. There is then only one circularly polarized wave transmitted in the plasma, which has the same sense of polarization as the incident wave. The results are almost identical to the isotropic case, except for the inclusion of the external magnetic field term in the expression for the refractive index.

In the present chapter we shall consider the reflection and the transmission of a linearly polarized electromagnetic wave, normally incident on a semi-infinite magneto-plasma, moving through a dielectric. The plasma is assumed to be moving in the direction normal to the boundary. Two different configurations of the static magnetic field are considered. In the first case, the direction of the longitudinal static magnetic field will be along the normal to the boundary. In the second case, the direction of the transverse static magnetic field will be parallel to the boundary. In general, for each case, two waves will be transmitted in the plasma. For the first case, both of the waves are circularly polarized with the opposite sense of rotation.

7.2 Formulation of the Problem

Let S' be the system of reference at rest in the moving plasma and S be the laboratory system of reference at rest in the dielectric. The corresponding coordinate axes of the two systems are assumed to be parallel. Let the plasma be moving along the x axis, where the plane x' = 0 is taken to be the boundary of the moving plasma. The primed quantities in the following analysis refer to the S' system, whereas the unprimed quantities refer to the S system. The geometry of the problem is shown in figure 8.

Let the dielectric constant of the dielectric in the S system be κ, and the uniform velocity of the plasma as observed in the laboratory system S be $-\bar{v}_0(-v_0,0,0)$. Thus in the S' system the dielectric

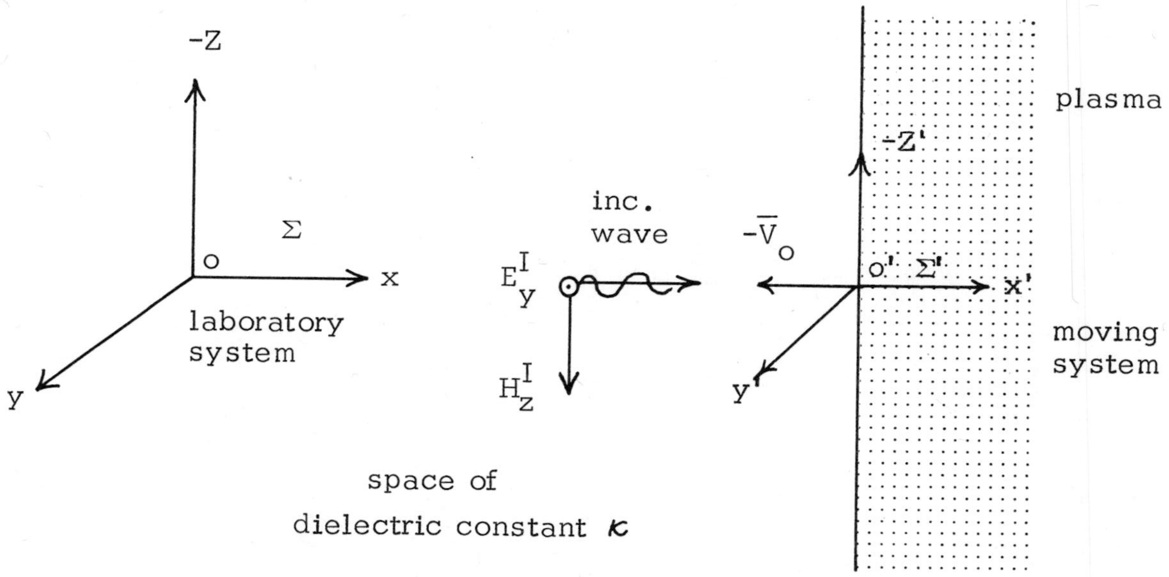

Figure 8. Normal Incidence on Semi-Infinite Magneto-Plasma Moving Through a Dielectric

moves with a uniform velocity $\bar{v}_o(v_o,0,0)$. Assuming harmonic time variation $e^{i\omega' t'}$ and the convective current model for the plasma, the Maxwell equations in S' may be written as follows:

$$\nabla' \times \bar{E}' = -i\omega'\bar{B}' \tag{7-1}$$

$$\nabla' \times \bar{H}' = \bar{J}' + i\omega'\bar{D}' \tag{7-2}$$

The constitutive relations in S', as given by Ohm's law [Brandstatter, 1963] and the Minkowski equations [Sommerfeld, 1964a] are:

$$\bar{J}' = \bar{\bar{\sigma}}' \cdot \bar{E}' \tag{7-3}$$

$$(1-\kappa\beta^2)\bar{D}' = \epsilon(1-\beta^2)\bar{E}' + \mu_o(\epsilon-\epsilon_o)\,[\bar{v}_o \times \bar{H}' - \epsilon\bar{v}_o(\bar{v}_o\cdot\bar{E}')] \tag{7-4}$$

$$(1-\kappa\beta^2)\bar{B}' = \mu_o(1-\beta^2)\bar{H}' - \mu_o(\epsilon-\epsilon_o)\,[\bar{v}_o \times \bar{E}' + \mu_o\bar{v}_o(\bar{v}_o\cdot\bar{H}')] \tag{7-5}$$

where $\bar{\bar{\sigma}}'$ is the conductivity tensor [Ratcliffe, 1959], $\beta = v_o/c$, and $\kappa = \epsilon/\epsilon_o$ is the dielectric constant.

Let the incident electromagnetic wave be propagating along the +x axis, the direction of propagation of the wave and the motion of the plasma thus being in opposite directions to each other. In order to determine the propagation properties of the waves transmitted in the plasma, the exponential space-time variation for those waves is assumed to be $e^{i(\omega' t' - k_o' n' x')}$, where $k_o' = \omega'\sqrt{\mu_o\epsilon_o} = \omega'/c$, and n' is the wave refractive index. Substituting the space variation and (7-3) - (7-5) in (7-1) - (7-2), one obtains:

$$n'\hat{x} \times \bar{E}' = \frac{c}{(1-\kappa\beta^2)}\,[\mu_o(1-\beta^2)\bar{H}' - \mu_o\epsilon_o(\kappa-1)(\bar{v}_o \times \bar{E}' + \mu_o v_o^2 H_x'\hat{x})] \tag{7-6}$$

$$n'\hat{x} \times \bar{H}' = \frac{\bar{\bar{\sigma}}' \cdot \bar{E}'}{-ik_o'} - \frac{c}{(1-\kappa\beta^2)}\,[\epsilon(1-\beta^2)\bar{E}' + \mu_o\epsilon_o(\kappa-1)(\bar{v}_o \times \bar{H}' - \epsilon v_o^2 E_x'\hat{x})] \tag{7-7}$$

Equations (7-6) - (7-7) give four simultaneous equations for the case when the static magnetic field $\bar{B}_o$

is along the longitudinal x axis, and five simultaneous equations when B_0 is along the transverse z axis. The condition that the corresponding determinants of the matrices of the coefficients of these equations be equal to zero for nontrivial solutions, yields the corresponding refractive index equations. These algebraic characteristic equations are found to be of the fourth order, and they will be discussed in the next two sections for the two configurations of the magnetic field. Two of the four refractive indices will represent the waves propagating in the +x' direction, and the other two refractive indices will represent the waves propagating in the -x' direction. Here we shall only be concerned with the waves in the +x' direction, and let n_j' for j = 1,2, be the two refractive indices corresponding to these two waves.

The refractive index as observed in the S system can be found by using (2-27) - (2-28) [Papas, 1965], as follows:

$$n_j = (n_j' - \beta)/(1 - n_j'\beta). \tag{7-8}$$

Let the incident electromagnetic wave be given in the S system by:

$$\bar{E}_I = \hat{y}E_y^I e^{i(\omega t - k_o Mx)} \tag{7-9}$$

$$\bar{H}_I = \hat{z}H_z^I e^{i(\omega t - k_o Mx)} \tag{7-10}$$

where $k_o = \omega\sqrt{\mu_o \epsilon_o}$ and $M = k/k_o = \sqrt{\epsilon/\epsilon_o} = \sqrt{\kappa}$, ω being the circular frequency of the incident wave ($\omega = \omega_I$). Let the electromagnetic reflected wave be given by:

$$\bar{E}_R = (\hat{y}E_y^R + \hat{z}E_z^R)e^{i(\omega_r t + k_r Mx)} \tag{7-11}$$

$$\bar{H}_R = (\hat{y}H_y^R + \hat{z}H_z^R)e^{i(\omega_r t + k_r Mx)} \tag{7-12}$$

where $k_r = \omega_r\sqrt{\mu_o \epsilon_o}$, ω_r being the frequency of the reflected wave. Using (7-9) - (7-12) in the Maxwell equation $\nabla \times \bar{E} = -i\omega\mu_o H$, one obtains:

$$\frac{E_y^I}{H_z^I} = -\frac{E_y^R}{H_z^R} = \frac{E_z^R}{H_y^R} = \frac{1}{M}\sqrt{\frac{\mu_o}{\epsilon_o}} = \frac{1}{M}\eta_o = \eta = \sqrt{\frac{\mu_o}{\epsilon}} \tag{7-13}$$

Let the incident and reflected waves be represented in the S' system by:

$$\bar{E}'_I = \hat{y}' E'^{I}_y e^{i(\omega' t' - k'_o M' x')} \tag{7-14}$$

$$\bar{H}'_I = \hat{z}' H'^{I}_z e^{i(\omega' t' - k'_o M' x')} \tag{7-15}$$

$$\bar{E}'_R = (\hat{y}' E'^{R}_y + \hat{z}' E'^{R}_z) e^{i(\omega'_r t' + k'_o M'_r x')} \tag{7-16}$$

$$\bar{H}'_R = (\hat{y}' H'^{R}_y + \hat{z}' H'^{R}_z) e^{i(\omega'_r t' + k'_o M'_r x')} \tag{7-17}$$

where $\omega'_r = \omega'$ and $k'_o = \omega'/c$. Using the property of phase invariance (2-24) in (7-9) - (7-10) and (7-14) - (7-15), one obtains from (2-25) - (2-28):

$$\omega' = \gamma\omega(1 + M\beta); \; M' = (M + \beta)/(1 + M\beta) \tag{7-18}$$

where $\gamma = (1 - \beta^2)^{-1/2}$.

The use of phase invariance (2-24) in (7-11) - (7-12) and (7-16) - (7-17) gives from (2-25) - (2-28):

$$\omega'_r = \gamma\omega_r(1 - M\beta); \; M'_r = (M - \beta)/(1 - M\beta) \tag{7-19}$$

Equating $\omega'_r = \omega'$ in (7-18) - (7-19) yields:

$$\omega_r = \omega\left(\frac{1 + M\beta}{1 - M\beta}\right) \tag{7-20}$$

Let us now consider the field transformations. The general laws of the field transformations are given by [Sommerfeld, 1964a]:

$$\bar{E}'_{\parallel} = (\bar{E} + \bar{v}_o \times \bar{B})_{\parallel} \; ; \; \bar{E}'_{\perp} = \gamma(\bar{E} + \bar{v}_o \times \bar{B})_{\perp} \tag{7-21}$$

$$H'_{\parallel} = (\bar{H} - \bar{v}_o \times \bar{D})_{\parallel} \; ; \; \bar{H}'_{\perp} = \gamma(\bar{H} - \bar{v}_o \times \bar{D})_{\perp} \tag{7-22}$$

Let us substitute the constitutive relations $\bar{B} = \mu_o\bar{H}$, $\bar{D} = \epsilon_o\kappa\bar{E}$ in the present convective current model into (7-21) - (7-22), where S' is moving with the velocity $(-v_o, 0, 0)$ with respect to S. Applying these results to (7-9) - (7-10), (7-14) - (7-15), and to (7-11) - (7-12), (7-16) - (7-17), one obtains:

$$E'^{I}_y = \gamma(1 + M\beta)E^{I}_y; \;\; \eta H'^{I}_z = \gamma(1 + M\beta)E^{I}_y \tag{7-23}$$

$$E_y'^R = \gamma(1 - M\beta)E_y^R; \quad \eta H_z'^R = -\gamma(1 - M\beta)E_y^R \tag{7-24}$$

$$E_z'^R = \gamma(1 - M\beta)E_z^R; \quad \eta H_y'^R = \gamma(1 - M\beta)E_z^R \tag{7-25}$$

From the relations (7-23) - (7-25) it is clear that:

$$\frac{E_y'^I}{H_z'^I} = \frac{-E_y'^R}{H_z'^R} = \frac{E_z'^R}{H_y'^R} = \eta = \sqrt{\frac{\mu_o}{\epsilon}} \tag{7-26}$$

The two transmitted waves in the S' frame of reference have their refractive indices as n_1' and n_2' for either direction of the static magnetic field. The analysis in the next two sections shows that we need to consider the TM and the TEM modes. We shall consider only the TM mode here, however, since the TEM mode can be found as a special case by putting $E_x' = 0$. The two transmitted waves can be represented in the S' frame as:

$$\overline{E}_T'^{(j)} = (\hat{x}'E_x'^{(j)} + \hat{y}'E_y'^{(j)} + \hat{z}'E_z'^{(j)})e^{i(\omega't'-k_o'n_j'x')} \tag{7-27}$$

$$\overline{H}_T'^{(j)} = (\hat{y}H_y'^{(j)} + \hat{z}'H_z'^{(j)})e^{i(\omega't' - k_o'n_j'x')} \tag{7-28}$$

and in the S frame as:

$$\overline{E}_T^{(j)} = (\hat{x}E_x^{(j)} + \hat{y}E_y^{(j)} + \hat{z}E_z^{(j)})e^{i(\omega_j t-k_j n_j x)} \tag{7-29}$$

$$\overline{H}_T^{(j)} = \hat{y}H_y^{(j)} \quad \hat{z}H_z^{(j)})e^{i(\omega_j t - k_j n_j x)} \tag{7-30}$$

where $j = 1,2$, $\omega' = \gamma\omega(1+M\beta)$, and $k_j = \omega_j/c$. Using the phase invariance (2-24) in (7-27) - (7-30), one obtains from (2-25) - (2-26):

$$\omega' = \gamma\omega_j(1+n_j\beta); \quad n_j' = \frac{n_j + \beta}{1+n_j\beta} \tag{7-31}$$

Using the relation $\omega' = \gamma\omega(1+M\beta)$ in (7-31), one has:

$$\omega_j = \frac{(1+M\beta)}{(1+n_j\beta)}\omega \tag{7-32}$$

From the Maxwell equation $\nabla \times \overline{E}^{(j)} = -i\omega_j\mu_o\overline{H}^{(j)}$ for $j = 1,2$, one finds:

$$H_x^{(j)} = 0; \quad \frac{E_y^{(j)}}{H_z^{(j)}} = - \frac{E_z^{(j)}}{H_y^{(j)}} = \frac{1}{n_j} \eta_o \tag{7-33}$$

Substituting (7-27) - (7-30) and (7-33) in (7-21) - (7-22), one obtains:

$$E_y'^{(j)} = \gamma(1+n_j \beta)E_y^{(j)}; \; H_z'^{(j)} = \gamma\sqrt{\frac{\epsilon_o}{\mu_o}}(n_j + \kappa\beta)E_y^{(j)} \tag{7-34}$$

$$E_z'^{(j)} = \gamma(1+n_j \beta)E_z^{(j)}; \; H_y'^{(j)} = -\gamma\sqrt{\frac{\epsilon_o}{\mu_o}}(n_j + \kappa\beta)E_z^{(j)} \tag{7-35}$$

$$E_x'^{(j)} = E_x^{(j)} \tag{7-36}$$

therefore, one has:

$$\frac{E_y'^{(j)}}{H_z'^{(j)}} = \sqrt{\frac{\mu_o}{\epsilon_o}} \; \frac{(1 + n_j \beta)}{(n_j + \kappa\beta)} = - \frac{E_z'^{(j)}}{H_y'^{(j)}}, \tag{7-37}$$

$$\frac{E_y'^{(j)}}{E_z'^{(j)}} = \frac{E_y^{(j)}}{E_z^{(j)}} \tag{7-38}$$

Substituting (7-8) in (7-37), one obtains:

$$\frac{E_y'^{(j)}}{H_z'^{(j)}} = \sqrt{\frac{\mu_o}{\epsilon_o}} \cdot \frac{(1 - \beta^2)}{n_j'(1 - \kappa\beta^2) - \beta(1 - \kappa)} = - \frac{E_z'^{(j)}}{H_y'^{(j)}} \tag{7-39}$$

In the S' frame of reference the following boundary conditions must be satisfied by the incident, reflected, and transmitted wave fields [Landau and Lifshitz, 1966]:

$$E_y'(+0) = E_y'(-0) \tag{7-40}$$

$$E_z'(+0) = E_z'(-0) \tag{7-41}$$

$$H_y'(+0) = H_y'(-0) \tag{7-42}$$

$$H_z'(+0) = H_z'(-0) \tag{7-43}$$

where the left-hand side refers to the total fields on the plasma side of the boundary, and the right-hand side refers to the total fields on the other side of it.

Since the transmitted waves for the two configurations of the static magnetic field are different, the formulas for the reflection and the transmission coefficients for the two cases will be different, and will be discussed in the next two sections.

7.3 The Longitudinal Static Magnetic Field

The conductivity tensor $\bar{\bar{\sigma}}'$ is given [Brandstatter, 1963; Ratcliffe, 1959; Budden, 1961] for the longitudinal static magnetic field $\bar{Y}' = Y'\hat{x}$, in the following form:

$$\bar{\bar{\sigma}}' = -i \frac{N'e'^2}{m'\omega'} \frac{1}{U'(U'^2 - Y'^2)} \begin{bmatrix} (U'^2 - Y'^2) & 0 & 0 \\ 0 & U'^2 & +iU'Y' \\ 0 & -iU'Y' & U'^2 \end{bmatrix} \qquad (7\text{-}44)$$

where we take (-e') as the electron charge and define:

$U' = 1 - iZ'$, $Z' = \nu'/\omega'$,

ν' = phenomenological collision frequency

$Y' = \omega'_{HL}/\omega'$, $\omega'_{HL} = eB'_o/m' = e\mu_o H'_o/m'$

ω'_{HL} = Gyromagnetic (cyclotron) frequency.

B'_o = magnitude of the applied static magnetic field along the x direction. Substituting (7-44) in (7-7) and expanding the resulting equation and (7-6), one obtains:

$$\left[n' - \frac{(1-\kappa)\beta}{(1-\kappa\beta^2)}\right] E'_z + \eta_o \frac{(1-\beta^2)}{(1-\kappa\beta^2)} H'_y = 0 \qquad (7\text{-}45)$$

$$\left[n' - \frac{(1-\kappa)\beta}{(1-\kappa\beta^2)}\right] E'_y - \eta_o \frac{(1-\beta^2)}{(1-\kappa\beta^2)} H'_z = 0 \qquad (7\text{-}46)$$

$$\frac{1}{\eta_o}\left[\frac{\kappa(1-\beta^2)}{(1-\kappa\beta^2)} - \frac{X'U'}{(U'^2 - Y'^2)}\right] E'_y - \frac{1}{\eta_o} \frac{iX'Y'}{(U'^2 - Y'^2)} E'_z - \left[n' - \frac{(1-\kappa)\beta}{(1-\kappa\beta^2)}\right] H'_z = 0 \qquad (7\text{-}47)$$

$$\frac{1}{\eta_0}\frac{iX'Y'}{(U'^2-Y'^2)}E'_y+\frac{1}{\eta_0}\left[\frac{\kappa(1-\beta^2)}{(1-\kappa\beta^2)}-\frac{X'U'}{(U'^2-Y'^2)}\right]E'_z+\left[n'-\frac{(1-\kappa)\beta}{(1-\kappa\beta^2)}\right]H'_y=0 \qquad (7\text{-}48)$$

where $X' = \omega_p'^2/\omega'^2$ and $\omega_p'^2 = N'e'^2/m'\epsilon_0$.

The condition for the non-trivial solution of (7-45) - (7-48) gives the refractive index:

$$n' = \frac{1}{(1-\kappa\beta^2)}\left[(1-\kappa)\beta \pm (1-\beta^2)\sqrt{\kappa - \frac{X'(1-\kappa\beta^2)}{(U'\pm Y')(1-\beta^2)}}\right] \qquad (7\text{-}49)$$

Here $H'_x = 0$ for $1 - \kappa\beta^2 \neq 0$, and $E'_x = 0$ for $X'/U' - \kappa \neq 0$, and thus under these conditions the waves are in TEM mode. The sign in front of the radical is chosen to be positive in the following, since for $\beta = 0$ and no plasma, it represents waves traveling in the positive x direction.

Substituting for n' from (7-49) into (7-39), one obtains after rearranging:

$$\frac{H_z'^{(j)}}{E_y'^{(j)}} = \frac{1}{\eta_0}\left[\kappa - \frac{X'(1-\kappa\beta^2)}{(U'-s_jY')(1-\beta^2)}\right]^{1/2} = \ell'_j/\eta_0 = \frac{-H_y'^{(j)}}{E_z'^{(j)}} \qquad (7\text{-}50)$$

where $s_1 = +1$, $s_2 = -1$.

Substituting (7-50) in (7-47) and using (7-38) in the result, gives:

$$E_y'^{(1)}/E_z'^{(1)} = E_y^{(1)}/E_z^{(1)} = +i \qquad (7\text{-}51)$$

$$E_y'^{(2)}/E_z'^{(2)} = E_y^{(2)}/E_z^{(2)} = -i \qquad (7\text{-}52)$$

The two transmitted waves are thus found to be circularly polarized in the opposite sense.

The plasma parameters used so far are the ones observed in the S' system of reference. It is desirable to transform these parameters to the S system. Using the transformations $\omega_p' = \omega_p$, $\nu' = \gamma\nu$, and $\omega'_{HL} = \gamma\omega_{HL}$ from section 2.4, and (7-18), one obtains:

$$X' = X/\gamma^2(1+M\beta)^2 \qquad (7\text{-}53)$$

$$U' = 1 - iZ/(1+M\beta) \qquad (7\text{-}54)$$

$$Y' = Y/(1+M\beta) \tag{7-55}$$

Substituting (7-53) - (7-55) in (7-49), one will obtain n_j' in terms of the plasma parameters observed in the laboratory frame; the result when substituted in (7-8) will give n_j in terms of these plasma parameters. The n_j appearing in what follows will refer to the n_j calculated in this manner. The symbol ℓ_j in the following corresponds to ℓ_j' after (7-53) - (7-55) have been substituted in (7-50).

Substituting now (7-14) - (7-17) and (7-27) - (7-28) in the boundary conditions (7-40) - (7-43) at $x' = 0$, one obtains:

$$E_y'^{(1)} + E_y'^{(2)} = E_y'^{R} + E_y'^{I} \tag{7-56}$$

$$E_z'^{(1)} + E_z'^{(2)} = E_z'^{R} \tag{7-57}$$

$$H_y'^{(1)} + H_y'^{(2)} = H_y'^{R} \tag{7-58}$$

$$H_z'^{(1)} + H_z'^{(2)} = H_z'^{R} + H_z'^{I} \tag{7-59}$$

Substituting (7-13), (7-23)- (7-26), (7-33), and (7-50) - (7-52) in (7-56) - (7-59), and cancelling out the common factor γ, one obtains:

$$\begin{bmatrix} (1+n_1\beta) & (1+n_2\beta) & -(1-M\beta) & 0 \\ -i(1+n_1\beta) & i(1+n_2\beta) & 0 & -(1-M\beta) \\ i\ell_1(1+n_1\beta) & -\ell_2(1+n_2\beta) & 0 & -M(1-M\beta) \\ \ell_1(1+n_1\beta) & \ell_2(1+n_2\beta) & M(1-M\beta) & 0 \end{bmatrix} \begin{bmatrix} E_y^{(1)} \\ E_y^{(2)} \\ E_y^{R} \\ E_z^{R} \end{bmatrix} = \begin{bmatrix} (1+M\beta) \\ 0 \\ 0 \\ M(1+M\beta) \end{bmatrix} E_y^{I} \tag{7-60}$$

Equation (7-60) is readily solved to give:

$$\frac{E_y^R}{E_y^I} = r_1 = \frac{(1+M\beta)}{(1-M\beta)}\left[1 - \frac{\ell_1}{(M+\ell_1)} - \frac{\ell_2}{(M+\ell_2)}\right] = \frac{1}{2}\left[\frac{(M-n_1)}{(M+n_1)} + \frac{(M-n_2)}{(M+n_2)}\right] \tag{7-61}$$

$$\frac{E_z^R}{E_y^I} = r_2 = \frac{i(1+M\beta)(\ell_1 - \ell_2)M}{(1-M\beta)(M+\ell_1)(M+\ell_2)} = \frac{iM(n_1 - n_2)}{(M+n_1)(M+n_2)} \tag{7-62}$$

$$\frac{E_y^{(1)}}{E_y^{I}} = t^{(1)} = \frac{M(1+M\beta)}{(M+\ell_1)(1+n_1\beta)} = \frac{M}{(M+n_1)} \tag{7-63}$$

$$\frac{E_y^{(2)}}{E_y^{I}} = t^{(2)} = \frac{M(1+M\beta)}{(M+\ell_2)(1+n_2\beta)} = \frac{M}{(M+n_2)} \tag{7-64}$$

For the particular case of $\kappa = 1$, $M = 1$, equations (7-61) - (7-64) agree with the results in (6-67) - (6-70).

For a circularly polarized incident electromagnetic wave, only one wave is transmitted into the plasma, with the same sense of circular polarization, and a circularly polarized electromagnetic wave is reflected from the plasma boundary, with the same sense of polarization as the incident wave, as discussed in detail in section 6.6. By comparing the results of (7-61) - (7-64) with (6-67) - (6-70), one is able to obtain, for the present case under consideration, the corresponding results for the circularly polarized waves discussed in section 6.6, in analogous manner to the ones given previously in (6-108) - (6-109), as follows:

$$r_{r\ell} = \frac{M - n_1}{M + n_1} \; ; \; t_{rr} = \frac{2M}{M + n_1} \tag{7-65}$$

$$r_{\ell r} = \frac{M - n_2}{M + n_2} \; ; \; t_{\ell\ell} = \frac{2M}{M + n_2} \tag{7-66}$$

where $r_{rr} = r_{\ell\ell} = t_{r\ell} = t_{\ell r} = 0$ in accordance with the discussion in section 6.6. Equations (7-65) - (7-66) are given for the two senses of circular polarization of the incident electromagnetic wave, and they can be shown to agree with the results by Fainberg and Tkalich [1959], after some algebraic manipulations.

For the particular case of no static magnetic field $Y' = Y = 0$ one has $n_1 = n_2 = n$, and from (7-65) and (7-66) one obtains $r_{r\ell} = r_{\ell r}$ and $t_{rr} = t_{\ell\ell}$. By substituting (7-53) - (7-55) in (7-49) and the result in (7-8), one will obtain, for the case of no magnetic field $Y = 0$ and no collisions $Z = 0$, the corresponding reflection coefficient for circularly polarized electromagnetic waves, incident on an isotropic, semi-infinite plasma, moving through a dielectric medium:

$$r_{r\ell} = r_{\ell r} = \frac{(M-n)}{(M+n)} = \frac{1 - [1-X(1-M\beta)/\kappa(1+M\beta)]^{1/2}}{1 + [1-X(1-M\beta)/\kappa(1+M\beta)]^{1/2}} \cdot \frac{(1+M\beta)}{(1-M\beta)} \tag{7-67}$$

where $\kappa = M^2$ and the negative sign has been chosen in the numerator in order to obtain no reflection for $X = 0$. Equation (7-67) agrees with the result by Lampert [1956].

The refractive index equation for the present analysis for electromagnetic waves propagating in an infinite, isotropic plasma, moving through an infinite dielectric medium, may be found for the case of no collisions by substituting in (7-49) $Y' = 0$, $Z' = 0$, $U' = 1$, and taking for the infinite media:

$$\omega_p' = \omega_p; \quad \omega' = \gamma\omega(1+n\beta); \quad X' = \frac{X}{\gamma^2(1+n\beta)^2} \tag{7-68}$$

The transformation (7-68) applies to the infinite plasma and corresponds to the parameters used in (6-39), while the transformation (7-53) applies to the bounded plasma and corresponds to the parameters used in (6-40). Substituting (7-68) and $n' = (n+\beta)/(1+n\beta)$ from (7-31) in (7-49), one obtains, after some algebraic manipulations, the following final result in the form:

$$n^2 = \kappa - X \tag{7-69}$$

Equation (7-69) agrees with the results obtained by Lampert [1956] and by Clemmow [1962], if one takes into account the change in notation used there. For the particular case of plasma moving in free space $\kappa = 1$, one obtains from (7-69) $n^2 = 1 - X$, which agrees with the previous result in (4-97), derived by Unz [1966a].

Let us define the power reflection coefficient for the sub-Cerenkov case $(1 - \kappa\beta^2 > 0)$, as in (6-76):

$$R = (\bar{E}^R \times \bar{H}^{R*}) / (\bar{E}^I \times \bar{H}^{I*}) = |r_1|^2 + |r_2|^2 \tag{7-70}$$

where * represents the complex conjugate. A similar definition for power transmission coefficient can be given only when the transmitted frequencies are real, namely, when the refractive index is real, as follows from (7-32). The computations are, therefore, performed for the power reflection coefficient only.

The asymptotic values of the reflection coefficients as $\beta \to 1/\sqrt{\kappa}$ are found to be:

$$r_1 = \frac{1}{\kappa}\left(\frac{\omega_p^2}{4\omega^2 - \omega\omega_H}\right) = \frac{1}{\kappa}\left(\frac{1}{4\Omega^2 - \Omega\Omega_H}\right) \tag{7-71}$$

$$r_2 = \frac{-i}{2\kappa\omega}\left(\frac{\omega_p^2\,\omega_H}{4\omega^2 - \omega_H^2}\right) = -\frac{i}{2\kappa\Omega}\left(\frac{\Omega_H}{4\Omega^2 - \Omega_H^2}\right) \tag{7-72}$$

where $\Omega = \omega/\omega_p$ and $\Omega_H = \omega_H/\omega_p$.

7.4 The Transverse Static Magnetic Field

Taking the static magnetic field direction along the z axis, the conductivity tensor $\bar{\bar{\sigma}}'$ is given [Ratcliffe, 1959; Budden, 1961; Brandstatter, 1963] for the transverse static magnetic field $\bar{Y}' = Y'\hat{z}$, in the following form:

$$\bar{\bar{\sigma}}' = -i\,\frac{N'e'^2}{m'\omega'}\,\frac{1}{U'(U'^2-Y'^2)}\begin{bmatrix} U'^2 & +iU'Y' & 0 \\ -iU'Y' & U'^2 & 0 \\ 0 & 0 & (U'^2-Y'^2) \end{bmatrix} \tag{7-73}$$

In the present section the transverse static magnetic field is taken to be stationary in the moving frame of reference (i.e. the moving plasma frame).

Substituting (7-73) in (7-7) and expanding the resulting equation and (7-6), one obtains:

$$\left[n' - \frac{(1-\kappa)\beta}{(1-\kappa\beta^2)}\right]E'_z + \eta_0\,\frac{(1-\beta^2)}{(1-\kappa\beta^2)}\,H'_y = 0 \tag{7-74}$$

$$\left[n' - \frac{(1-\kappa)\beta}{(1-\kappa\beta^2)}\right]E'_y + \eta_0\,\frac{(1-\beta^2)}{(1-\kappa\beta^2)}\,H'_z = 0 \tag{7-75}$$

$$\left[\kappa - \frac{X'U'}{(U'^2 - Y'^2)}\right] E'_x - \frac{iX'Y'}{(U'^2 - Y'^2)} E'_y = 0 \qquad (7\text{-}76)$$

$$\frac{1}{\eta_0}\left[+ \frac{iX'Y'}{(U'^2 - Y'^2)}\right] E'_x + \frac{1}{\eta_0}\left[\frac{\kappa(1 - \beta^2)}{(1 - \kappa\beta^2)} - \frac{X'U'}{(U'^2 - Y'^2)}\right] E'_y - \\ - \left[n' - \frac{(1 - \kappa)\beta}{(1 - \kappa\beta^2)}\right] H'_z = 0 \qquad (7\text{-}77)$$

$$\frac{1}{\eta_0}\left[\frac{(1 - \beta^2)\kappa}{(1-\kappa\beta^2)} - \frac{X'}{U'}\right] E'_z + \left[n' - \frac{(1 - \kappa)\beta}{(1 - \kappa\beta^2)}\right] H'_y = 0 \qquad (7\text{-}78)$$

The condition for a non-trivial solution of (7-74) and (7-78) gives:

$$n' = \frac{1}{(1 - \kappa\beta^2)}\left\{(1 - \kappa)\beta \pm (1 - \beta^2)\left[\kappa - \frac{X'}{U'}\,\frac{(1 - \kappa\beta^2)}{(1 - \beta^2)}\right]^{1/2}\right\} \qquad (7\text{-}79)$$

The condition for a non-trivial solution of (7-75) - (7-77) gives:

$$n' = \frac{(1 - \kappa)\beta}{(1 - \kappa\beta^2)} \pm \left\{\frac{(1 - \beta^2)}{(1 - \kappa\beta^2)\left[\kappa - \frac{X'U'}{(U'^2 - Y'^2)}\right]} \cdot \left[\frac{X'^2}{(U'^2 - Y'^2)} - \right.\right. \\ \left.\left. - \frac{X'U'}{(U'^2 - Y'^2)}\left(\kappa + \frac{\kappa(1 - \beta^2)}{(1 - \kappa\beta^2)}\right) + \frac{\kappa^2(1 - \beta^2)}{(1 - \kappa\beta^2)}\right]\right\}^{1/2} \qquad (7\text{-}80)$$

In (7-79) and (7-80) the plus or minus sign in front of the square root term is chosen as discussed in what follows (7-49). Equation (7-79) corresponds to the ordinary mode, and (7-80) corresponds to the extraordinary mode. Here $H'_x = 0$ if $1 - \kappa\beta^2 \neq 0$, so the waves would be in the TM mode.

Substituting for n' from (7-79) and from (7-80) into (7-39), one obtains:

$$\frac{H'_z(1)}{E'_y(1)} = \frac{1}{\eta_0}\left[\kappa - \frac{X'}{U'}\,\frac{(1 - \kappa\beta^2)}{(1 - \beta^2)}\right]^{1/2} = \frac{-H'_y(1)}{E'_z(1)} = m'_1/\eta_0 \qquad (7\text{-}81)$$

$$\frac{H_z'^{(2)}}{E_y'^{(2)}} = \frac{1}{\eta_0}\left\{\frac{(1-\kappa\beta^2)}{(1-\beta^2)\left[\kappa - \frac{X'U'}{(U'^2-Y'^2)}\right]}\left[\frac{X'^2}{(U'^2-Y'^2} - \right.\right.$$

$$\left.\left. - \frac{X'U'}{(U'^2-Y'^2)}\left(\kappa + \frac{(1-\beta^2)}{(1-\kappa\beta^2)}\right) + \frac{\kappa^2(1-\beta^2)}{(1-\kappa\beta^2)}\right]\right\}^{1/2} = \frac{-H_y'^{(2)}}{E_z'^{(2)}} = m_2'/\eta_0 \qquad (7\text{-}82)$$

Substituting (7-79) and (7-81) into (7-77) gives:

$$\left[\frac{iX'\Omega'}{U'^2-\Omega'^2}\right] E_x' + \left[\frac{X'}{U'} - \frac{X'U'}{(U'^2-\Omega'^2)}\right] E_y' = 0$$

When this expression is compared with (7-76) it leads to $E_y'^{(1)} = E_x'^{(1)} = H_z'^{(1)} = 0$, unless the system has particular parameters, which will satisfy the equation to be obtained by equating the determinant of the coefficient matrix to zero, and we assume that this is not the case. Thus, the ordinary wave has components $E_z'^{(1)}$ and $H_y'^{(1)}$. By a similar substitution of (7-80) and (7-82) in (7-78), one gets $H_y'^{(2)} = E_z'^{(2)} = 0$. Thus, the extraordinary mode has components $E_x'^{(2)}$, $E_y'^{(2)}$, and $H_z'^{(2)}$. However, $E_x'^{(2)}$ and $H_z'^{(2)}$ can be written in terms of $E_y'^{(2)}$, using (7-76) and (7-82), respectively.

The plasma parameters X' and U' in this case transform as in (7-53) and (7-54), but since for the transverse magnetic field one has from (2-41) $\omega_{HT}' = \gamma^2\omega_{HT}$, we shall have for the present case, instead of (7-55), the following transformation:

$$Y' = \gamma Y/(1+M\beta) \qquad (7\text{-}83)$$

Substituting (7-53), (7-54), and (7-83) in (7-79) and in (7-80) will give n_j' in terms of the laboratory plasma parameters, and also the n_j, by use of (7-8). The letter m_j in the following corresponds to m_j', after (7-53), (7-54), and (7-83) have been substituted in (7-81) - (7-82).

Substituting (7-14) - (7-17) and (7-27) - (7-28) in the boundary conditions at x' = 0 given in (7-40) - (7-43), with due consideration to the above discussion, one obtains:

$$E_y'^{(2)} = E_y'^{R} + E_y'^{I} \tag{7-84}$$

$$E_z'^{(1)} = E_z'^{R} \tag{7-85}$$

$$H_y'^{(1)} = H_y'^{R} \tag{7-86}$$

$$H_z'^{(2)} = H_z'^{R} + H_z'^{I} \tag{7-87}$$

Substituting (7-13), (7-23) - (7-26), (7-33), and (7-81) - (7-82) into (7-84) - (7-87), and cancelling out the common factor γ, one obtains:

$$\begin{bmatrix} 0 & (1+n_2\beta) & -(1-M\beta) & 0 \\ (1+n_1\beta) & 0 & 0 & -(1-M\beta) \\ -m_1(1+n_1\beta) & 0 & 0 & -M(1-M\beta) \\ 0 & m_2(1+n_2\beta) & M(1-M\beta) & 0 \end{bmatrix} \begin{bmatrix} E_z^{(1)} \\ E_y^{(2)} \\ E_y^{R} \\ E_z^{R} \end{bmatrix} = \begin{bmatrix} (1+M\beta) \\ 0 \\ 0 \\ M(1+M\beta) \end{bmatrix} E_y^{I} \tag{7-88}$$

which gives the following results:

$$E_z^{R} = E_z^{(1)} = 0 \tag{7-89}$$

$$E_y^{(2)}/E_y^{I} = 2M(1+M\beta)/(M+m_2)(1+n_2\beta) \tag{7-90}$$

$$E_y^{R}/E_y^{I} = (1+M\beta)(M-m_2)/(M+m_2)(1-M\beta). \tag{7-91}$$

The power reflection coefficient can be defined, as in (7-70), to be:

$$R = |E_y^{R}/E_y^{I}|^2 \tag{7-92}$$

For the case of $\beta = 0$ for non-moving magnetoplasma, the results (7-89) and (7-91) agree with Budden [1961], while for the case of no static magnetic field, the results satisfy (7-67) for the circular polarization case.

7.5 Numerical Results and Discussion

The power reflection coefficients for the two cases under study in the present chapter were computed for zero collision frequency, and the results are plotted in Figures 9-12. Figures 9-10 are for the case of a longitudinal magneto-static field along the x-axis, and Figures 11 and 12 are for the case of a transverse magneto-static field along the z axis. The dielectric constant was chosen to be $\kappa = 9$ for sample calculations. Only the sub-Cerenkov case $(1 - \kappa\beta^2 > 0)$ was considered. For the Cerenkov case $(1 - \kappa\beta^2 < 0)$ the reflection coefficient will be zero.

Figure 9 shows curves of the power reflection coefficient R, versus the plasma relative velocity β. The graphs were plotted for $\Omega = 0.1$ and $\Omega_H = 0, 0.05, 0.2$, and 0.5. The subscripts L and T in Ω_H are dropped in the present discussion and are implied for the case considered. It is observed that $R < 1$ for $\beta < 0$, namely, the reflection coefficient is smaller than unity when the semi-infinite plasma is moving away from the source. For the isotropic case $\Omega_H = 0$, it is found that $R > 1$ for $\beta > 0$, with a maximum occuring at $\beta = 0.278$, and then down asymptotically up to $\beta = 0.333$. For $\Omega_H = 0.05$ it is found that the curve is identical with that for $\Omega_H = 0$ between $\beta = -0.13$ and $\beta = 0.266$, thus showing no effect of this small value of the external static magnetic field on the power reflection coefficient for this range of velocity. However, between $\beta = 0.266$ and $\beta = 0.333$, R has two maxima, but the larger of the two values is slightly less than the maximum value for the isotropic case. The most interesting curve is for $\Omega_H = 0.2$. This is due to the fact that the denominators in (7-71) and (7-72) become zero at $\Omega_H = 0.2$ and $\Omega = 0.1$, thus showing large reflection coefficients as $\beta \to 1/\sqrt{\kappa}$. For increasing values of Ω_H, the reflection coefficient decreases. It should be noted that for $\Omega_H \neq 0$, two maxima occur for the reflection coefficient.

Figure 10 shows curves of R versus the normalized wave frequency $\Omega = \omega/\omega_p$ for $\beta = 0.2$ and $\Omega_H = 0.$, 0.05, 0.2, and 0.5. At $\Omega_H = 0.$, the reflection coefficient is maximum at $R = 16.0$ for $\Omega < 0.16$, and R decreases monotonically for $\Omega > 0.16$. Inclusion of the magnetic field term shows more than one maximum; however, these maxima have smaller values compared to the maximum for the isotropic case. For each Ω_H there is a critical wave frequency, so that for any frequency greater than that, the reflection coefficient is higher

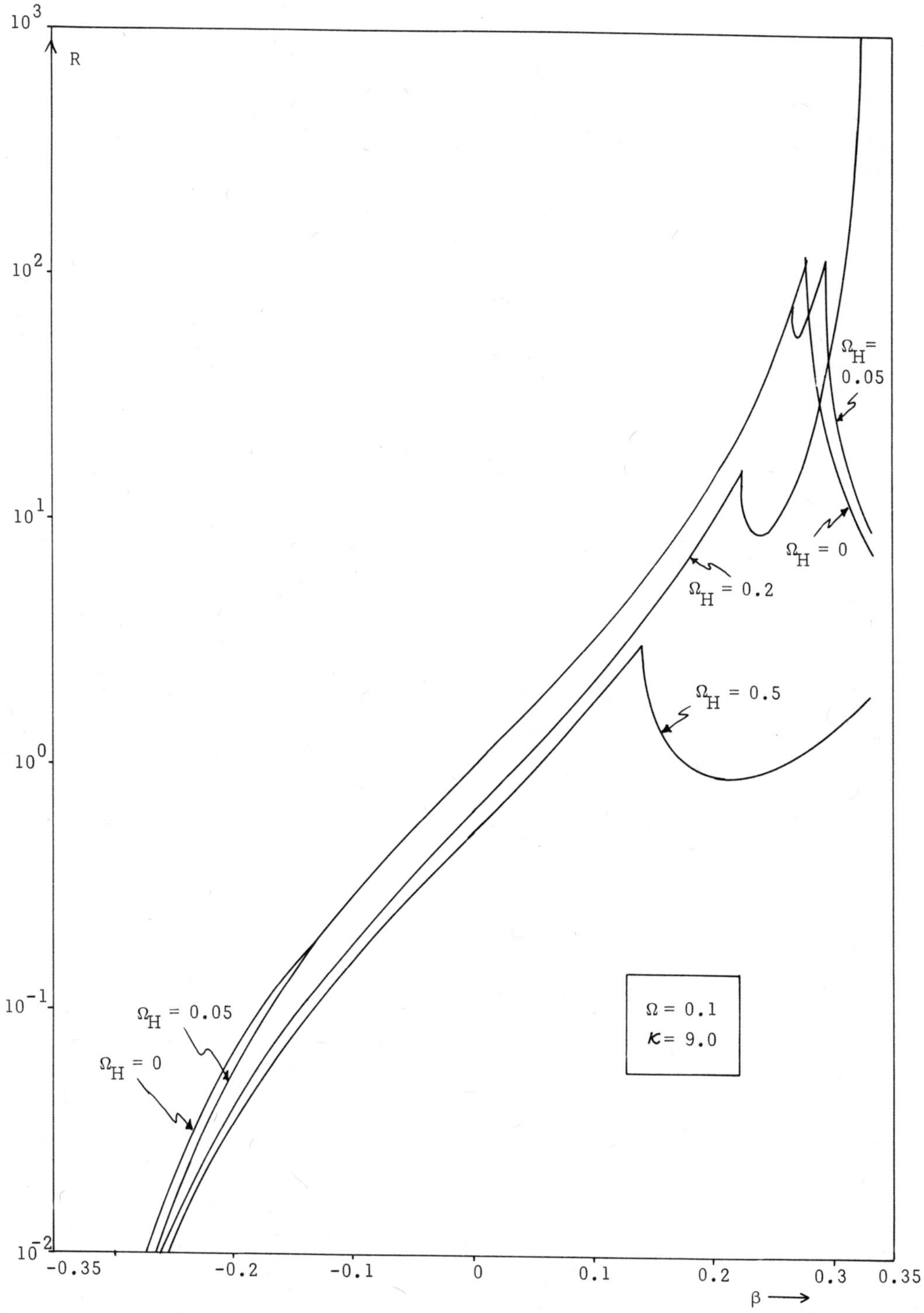

Figure 9. Reflection Coefficient vs. Normalized Velocity for Longitudinal Magnetic Field

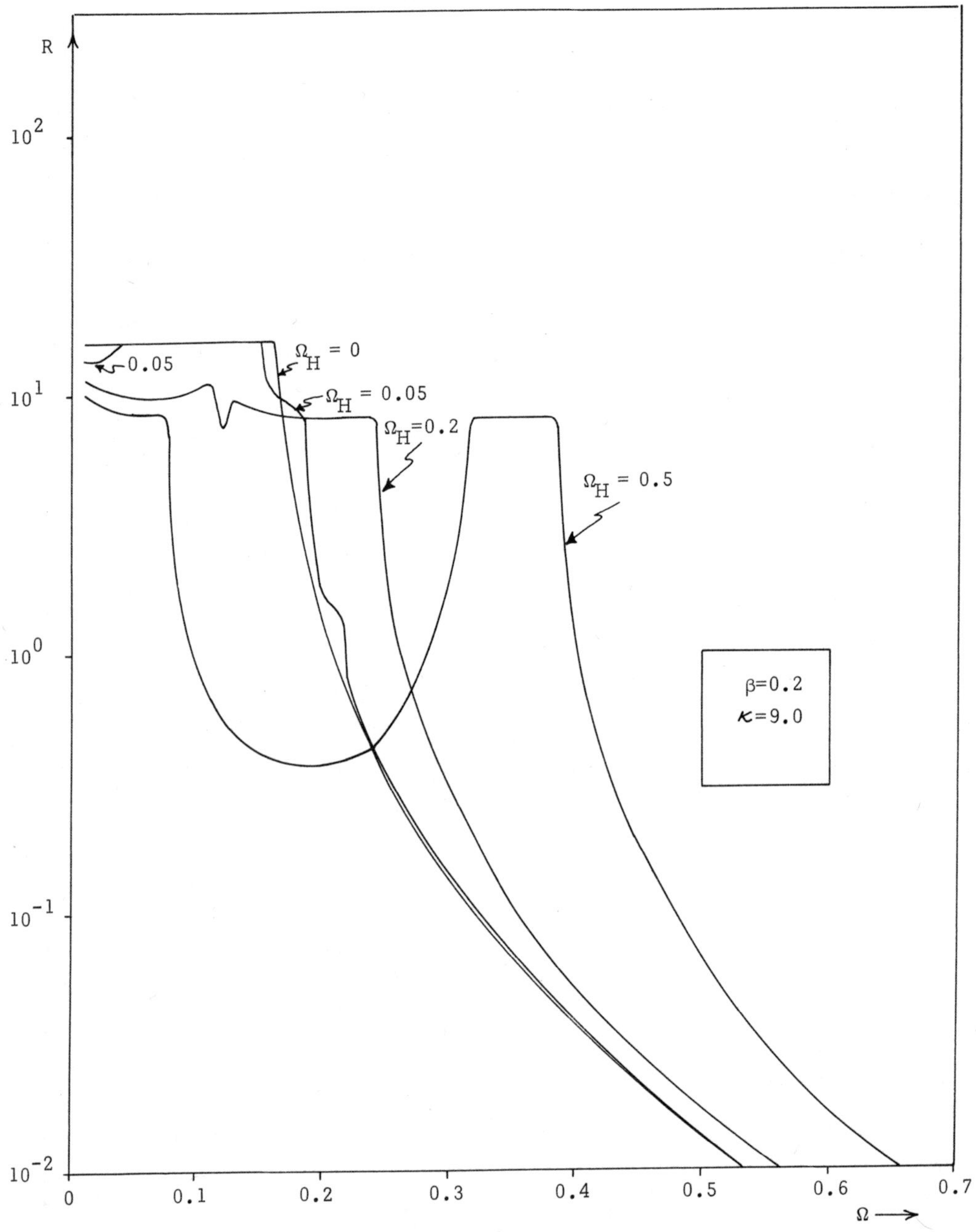

Figure 10. Reflection Coefficient vs. Normalized Incident Wave Frequency for Longitudinal Magnetic Field

than the corresponding values for the isotropic case. Consider the frequency region $0.32 < \Omega < 0.38$. The reflection coefficient for $\Omega_H \neq 0$ is larger than the value for the isotropic case $\Omega_H = 0$ in this region. Also, for $\Omega_H = 0.5$ there occurs a maximum in this region. This region, therefore, seems interesting for a plot of R vs. β. Calculations for R vs. β for $\Omega = 0.355$, and $\Omega_H = 0.5$ show essentially the same characteristics as those for $\Omega = 0.1$, and $\Omega_H = 0.5$, but with a maximum $R = 22.6$ at $\beta = 0.247$.

Figure 11 shows curves of R versus β for $\Omega_H = 0, 0.05, 0.2, 0.5$, but this time $\bar{\Omega}_H = \hat{z}\Omega_H$. It is interesting to note that the graphs for $\Omega_H \neq 0$ are identical with the one for $\Omega_H = 0$ for $\beta < \beta_{max}$, where β_{max} denotes velocity of the plasma at which R is maximum. Furthermore, this maximum is smaller than the maximum for $\Omega_H = 0$, and it occurs at lower velocity for increasing values of Ω_H.

Figure 12 shows curves of R versus Ω for $\beta = 0.2$ and $\Omega_H = 0., 0.05$, and 0.2. The maximum reflection coefficient occurs for the isotropic case at $R = 16.0$ for $\Omega < 0.16$. For $\Omega_H \neq 0$ the reflection coefficient oscillates in the frequency range $\Omega = 0.1$-0.3. As observed in Figure 10, there is a critical frequency beyond which the reflection coefficient for $\Omega_H \neq 0$ is greater than that for $\Omega_H = 0$.

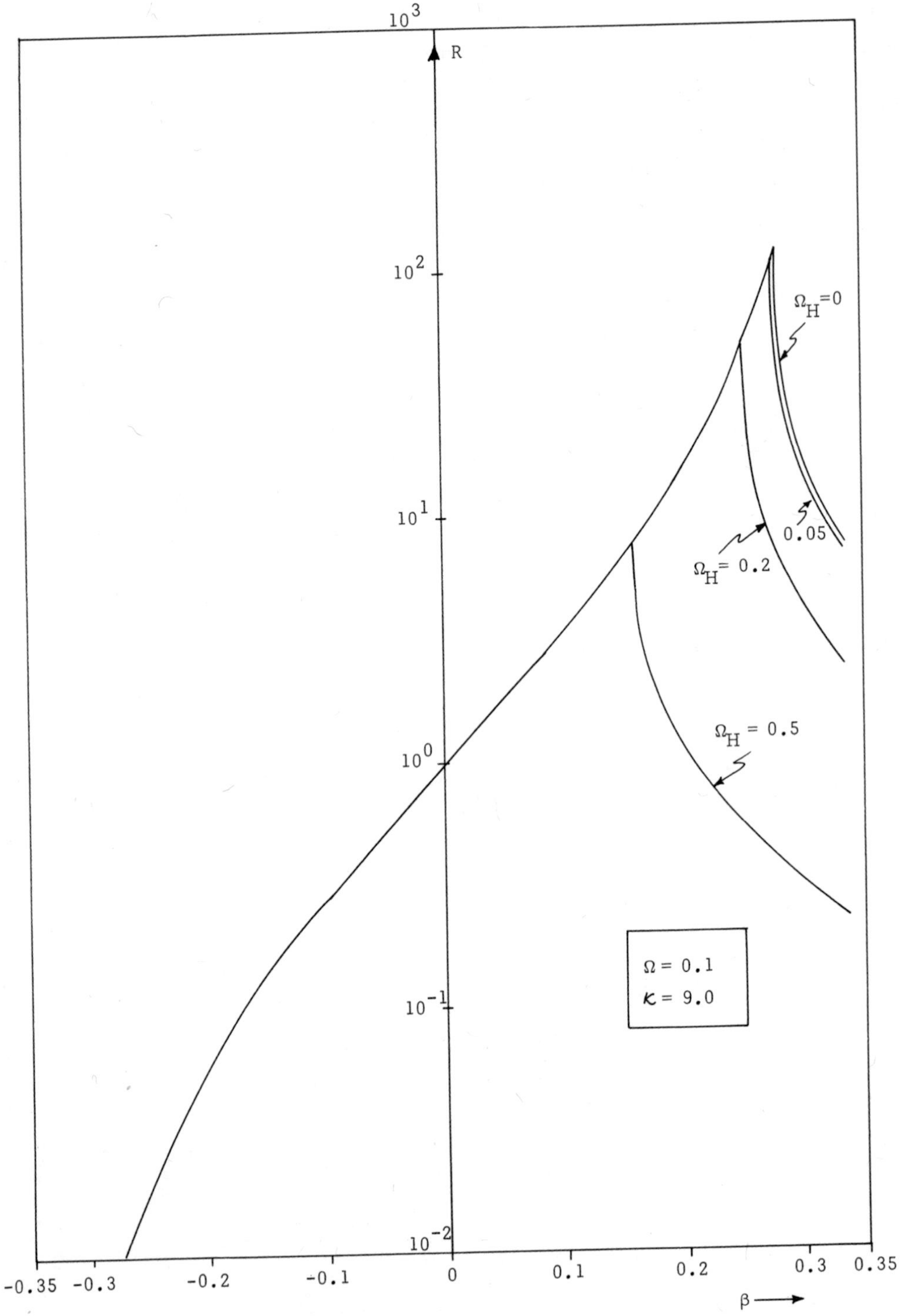

Figure 11. Reflection Coefficient vs. Normalized Velocity for Transverse Magnetic Field

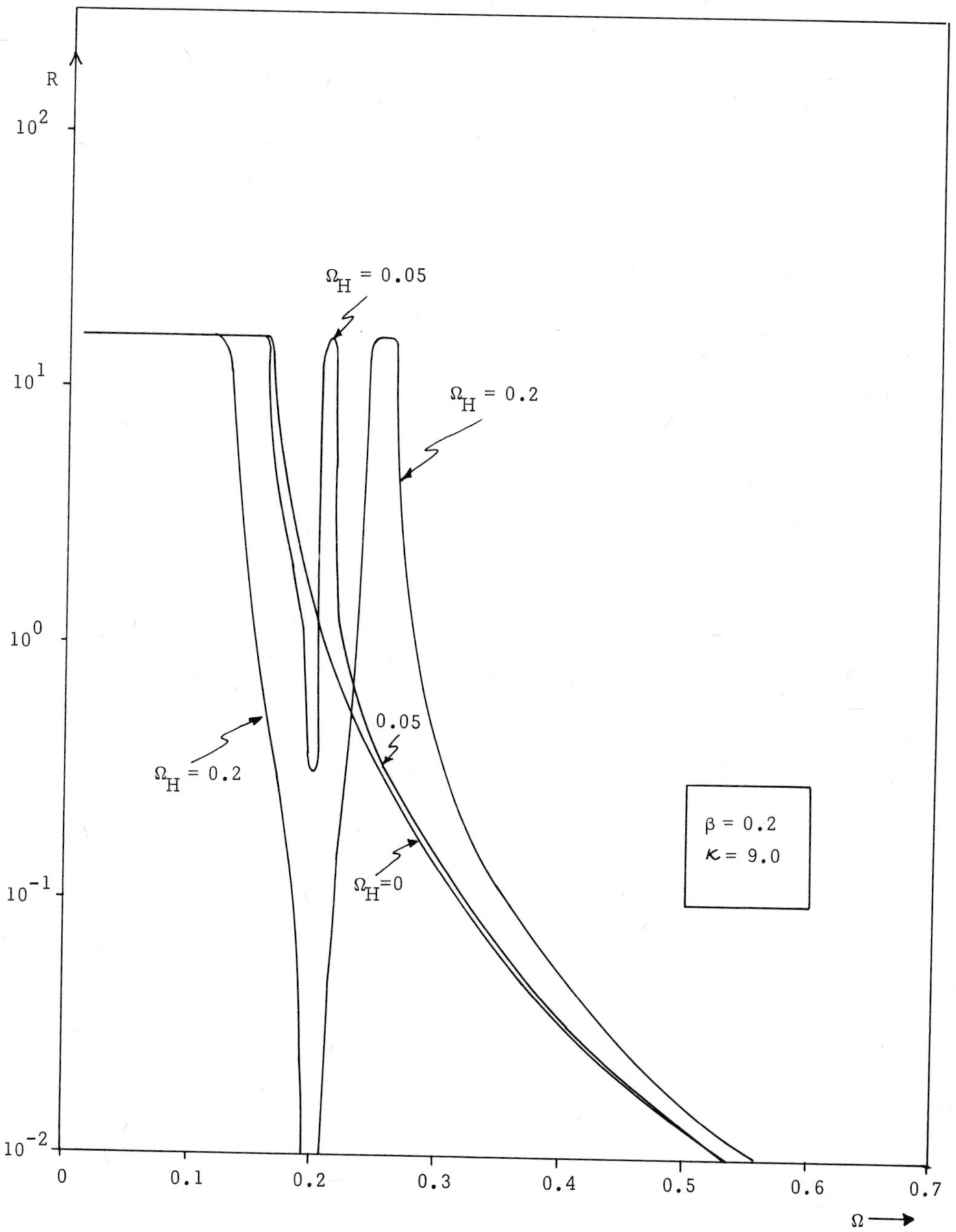

Figure 12. Reflection Coefficient vs. Normalized Incident Wave Frequency for Transverse Magnetic Field

VIII

NORMAL INCIDENCE ON A TEMPERATE MAGNETO-PLASMA SLAB MOVING THROUGH FREE SPACE

8.1 Introduction

The present chapter treats the reflection and the transmission of a linearly and a circularly polarized, normally incident, plane electromagnetic wave, propagating through a plasma slab uniformly moving along a magneto-static field which is perpendicular to the slab boundaries [Chawla and Unz, 1969b]. The reflection and the transmission of electromagnetic waves by a moving dielectric slab and a plasma slab, in the absence of the magneto-static field, have been considered by Yeh and Casey [1966] and by Yeh [1966], respectively. The aim of this chapter is to investigate the effect of the presence of the magneto-static field on the reflected and the transmitted waves from the plasma slab. The magneto-static field is assumed to be normal to the boundary. In the magneto-plasma slab, four waves are excited, two traveling towards one interface, and the other two traveling towards the second interface, as compared to the case of isotropic plasma, where only two waves are excited in the slab, with one traveling towards each interface. The particular case of moving, semi-infinite, isotropic plasma is considered, and the results are found to agree with previous results [Yeh, 1966].

The physical models of the present problem may be considered as a radiating source moving relatively to a stationary plasma slab, or a plasma slab moving relatively to a laboratory, where the radiating source is stationary; thus, in both cases, the slab is assumed to be moving with respect to the radiating source. Let us denote the coordinate system in which the radiating source is stationary as the laboratory system S, and the coordinate system which is embedded in the moving plasma and is stationary with respect to the moving plasma slab, as the rest system of the plasma S'. The corresponding coordinates (x, y, z) of the laboratory system S and the coordinates (x', y', z') of the rest system of the plasma S', are parallel to each other, and the geometry of the problem is shown in figure 13.

In the present chapter the reflected, transmitted, and plasma waves are determined for the case of a linearly or circularly polarized electromagnetic wave, normally incident on a magneto-plasma slab, moving along a magneto-static field in the longitudinal direction. The direction of the motion and the magneto-static field are both normal to the slab boundaries. The case of a linearly polarized incident electromagnetic wave is solved by extending the relativistic solution given in section 6.3 to the present case of plasma slab. The boundary conditions are applied, and the problem is solved, in the rest frame of reference S' of the plasma slab; and then relativistic transformations for the fields and the plasma parameters are applied in order to find the reflection and the transmission coefficients observed in the laboratory frame of reference. The same problem may be solved by extending the non-relativistic solution given in section 6.2 to the present case, and the results are found to be identical. The power reflection and power transmission coefficients of the linearly polarized incident electromagnetic waves through the magneto-plasma slab are evaluated numerically for particular cases, with the longitudinal magneto-static field as a parameter, and are discussed in detail.

The case of a circularly polarized incident electromagnetic wave is solved by extending the non-relativistic solution given in sections 6.2 and 6.6 to the present case of plasma slab. The same problem of circular polarization may be solved by extending the relativistic solution in section 8.3 for the linearly polarized case, and the results are found to be identical.

For the linearly polarized, normally incident, plane electromagnetic wave, on a plasma slab, moving longitudinally along a static magnetic field, one finds the final result, for both the relativistic and non-relativistic solution, in the form of an 8 x 8 matrix relationship between the four plasma wave fields, the reflected waves and the transmitted waves, and the given incident linearly polarized wave. For the case of the

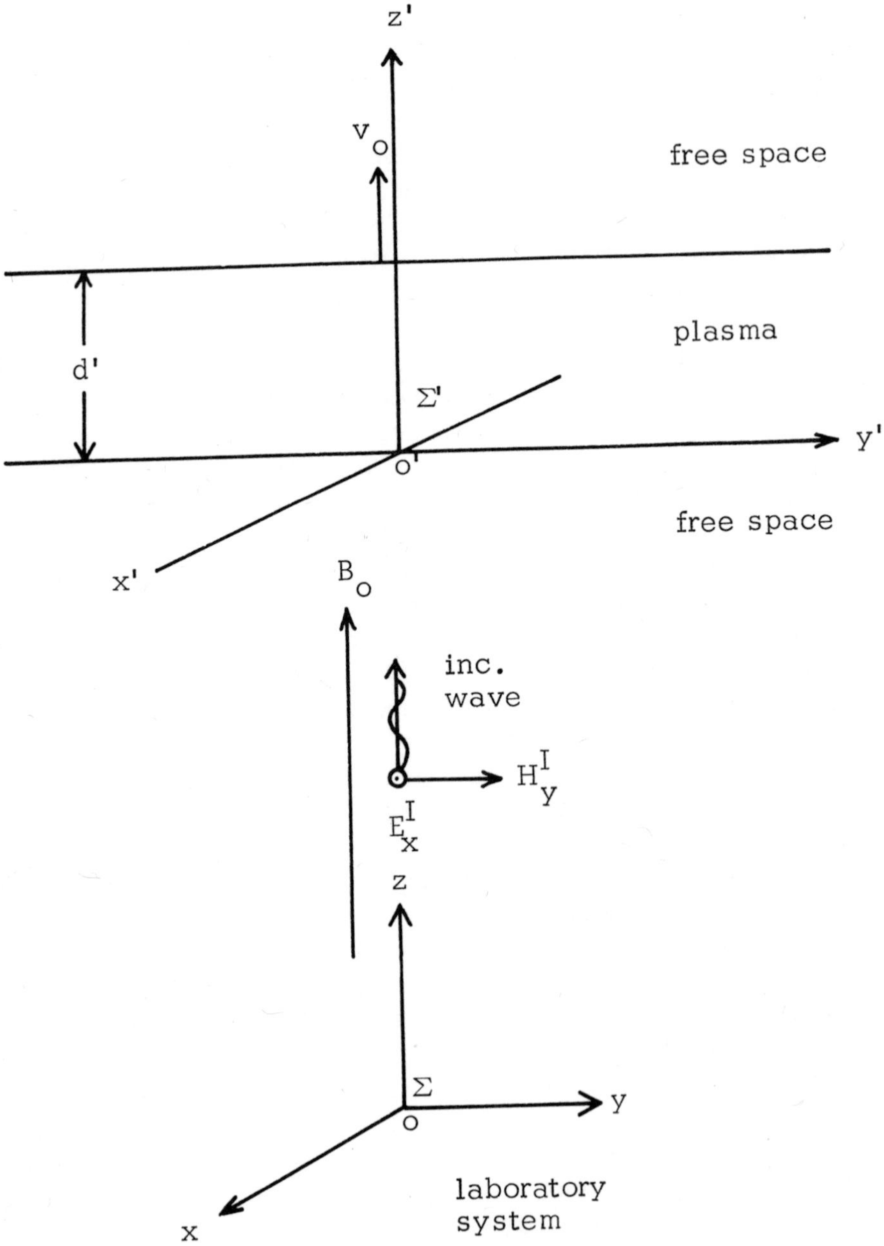

Figure 13. Normal Incidence on Moving Temperate Magneto-Plasma Slab

circularly polarized incident wave, in either the clockwise or the counterclockwise sense, one can separate the above 8 x 8 matrix into two independent 4 x 4 matrices, which can be solved explicitly for the corresponding reflection and transmission coefficients, separately for each one of circular polarization senses.

8.2 The Relativistic Solution

A linearly polarized, plane electromagnetic wave, normally incident on the plasma slab, described in the laboratory system, is first transformed to the rest system of the slab, where the boundary conditions are applied to the incident, reflected, plasma and transmitted waves. Then using the plasma parameter transformations and the wave-field transformations, the results for the reflected and the transmitted waves in reference to the laboratory system are expressed through an 8 x 8 matrix relation in terms of the incident wave electric field.

Consider a plane wave excited in the laboratory system S, propagating in free space in the +z direction. Let the field components of this wave vary exponentially as $e^{i(\omega t - k_o z)}$, where $\omega = \omega_I$ is the incident wave frequency and $k_o = \omega/c$ is the free space wave number. The unprimed quantities will correspond to the laboratory system S, and the primed quantities to the rest system S'. Let the rest system of the slab, denoted by S', be such that its coordinate axes (x', y', z') are parallel to the coordinate axes (x, y, z) of the laboratory system S. Let d' be the width of the plasma slab in S', bounded by the planes z' = 0 and z' = d' (Figure 13). Let the slab be moving with a uniform velocity $\bar{v}_0(0,0,v_0)$ along the +z direction, relative to the system S. The magneto-static field is also assumed to be along the +z direction. A negative velocity would then mean that the slab is moving towards the source of radiation.

Let the incident, reflected and transmitted waves in free space, observed in the rest system S', be represented as follows; for the incident wave:

$$\bar{E}'_{inc} = \hat{x}E^I_{x'}e^{i(\omega' t' - k'_o z')} \tag{8-1}$$

$$\bar{H}'_{inc} = \hat{y}H^I_{y'}e^{i(\omega' t' - k'_o z')} \tag{8-2}$$

where $E_{x'}^{I}$ and $H_{y'}^{I}$ are constants. For the reflected wave:

$$\bar{E}'_{ref} = (\hat{x}E_{x'}^{R} + \hat{y}E_{y'}^{R})e^{i(\omega' t' + k'_o z')} \tag{8-3}$$

$$\bar{H}'_{ref} = (\hat{x}H_{x'}^{R} + \hat{y}H_{y'}^{R})e^{i(\omega' t' + k'_o z')} \tag{8-4}$$

where $E_{x'}^{R}$, $E_{y'}^{R}$, $H_{x'}^{R}$ and $H_{y'}^{R}$ are constants. For the transmitted wave:

$$\bar{E}'_{tr} = (\hat{x}E_{x'}^{T} + \hat{y}E_{y'}^{T})e^{i(\omega' t' - k'_o z')} \tag{8-5}$$

$$\bar{H}'_{tr} = (\hat{x}H_{x'}^{T} + \hat{y}H_{y'}^{T})e^{i(\omega' t' - k'_o z')} \tag{8-6}$$

where $E_{x'}^{T}$, $E_{y'}^{T}$, $H_{x'}^{T}$ and $H_{y'}^{T}$ are constants. Equations (8-1) - (8-6) satisfy the Maxwell equations in the S' system under the conditions:

$$\frac{E_{x'}^{I}}{H_{y'}^{I}} = \frac{-E_{x'}^{R}}{H_{y'}^{R}} = \frac{E_{y'}^{R}}{H_{x'}^{R}} = \frac{E_{x'}^{T}}{H_{y'}^{T}} = \frac{-E_{y'}^{T}}{H_{x'}^{T}} = \sqrt{\frac{\mu_o}{\epsilon_o}} = \eta_o \tag{8-7}$$

In the laboratory system S one has similar expressions as (8-1) - (8-7), except that there are no primes and that for the reflected and the transmitted waves the corresponding frequencies and wave numbers are ω_R, ω_T, k_R, k_T instead of ω', k'_o.

The relations between the primed and unprimed electromagnetic fields are obtained through the Lorentz transformations in (2-13) and (2-16) as follows:

$$E_{x'}^{I} = \gamma(1 - \beta_L)E_x^{I}; \quad H_{y'}^{I} = \gamma(1 - \beta_L)E_x^{I}/\eta_o \tag{8-8}$$

$$E_{x'}^{R} = \gamma(1 + \beta_L)E_x^{R}; \quad E_{y'}^{R} = \gamma(1 + \beta_L)E_y^{R} \tag{8-9}$$

$$E_{x'}^{T} = \gamma(1 - \beta_L)E_x^{T}; \quad E_{y'}^{T} = \gamma(1 - \beta_L)E_y^{T} \tag{8-10}$$

One also obtains from the Lorentz transformations (2-25)-(2-26):

$$\omega' = \gamma\omega(1 - \beta_L); \quad k'_o = \gamma k_o(1 - \beta_L) \tag{8-11}$$

$$\omega' = \gamma\omega_R(1 + \beta_L); \quad k'_o = \gamma k_R(1 + \beta_L) \tag{8-12}$$

$$\omega' = \gamma\omega_T(1 - \beta_L): \quad k'_o = \gamma k_T(1 - \beta_L) \tag{8-13}$$

where $\beta_L = v_0/c$ and $\gamma = (1 - \beta_L^2)^{-1/2}$. From (8-11) - (8-13), one obtains:

$$\omega_R = \omega(1 - \beta_L)/(1+\beta_L); \quad \omega_T = \omega; \quad \omega = \omega_I \qquad (8\text{-}14)$$

$$k_R = k_0(1 - \beta_L)/(1+\beta_L); \quad k_T = k_0; \quad k_0 = \omega/c = \omega_I/c \qquad (8\text{-}15)$$

One finds that the reflected wave has a Doppler effect, whereas the transmitted wave does not; the effect of the moving slab on the reflected and the transmitted wave frequencies thus is independent of the medium of the slab. However, the polarizations of the reflected and the transmitted waves are now different from that of the incident wave for the general case of a moving magneto-plasma slab.

The refractive indices of the characteristic waves excited in the magneto-plasma slab in the rest system S' are given by the Appleton-Hartree equation (6-34), and its corresponding transformation to the laboratory system S, with respect to the incident wave frequency $\omega = \omega_I$, is given in (6-40).

In the derivation of (6-34), the field quantities in the plasma were assumed to vary exponentially as $e^{i(\omega' t' - k_0' n' z')}$, and for the non-resonant case under study it was found that $E_z' = 0$. Thus, in the rest system of the slab S' there are four waves given by (6-34), two of them traveling towards the interface $z' = 0$ and the other two towards the other interface $z' = d'$. The refractive index equation (6-34) was transformed to the laboratory system S, using the plasma parameters transformations of section 2.4 and (6-36), and is given by (6-40). Let the four roots of (6-34) be n_1', n_2', n_3' and n_4' such that $n_1' = -n_3'$ and $n_2' = -n_4'$. Let n_j, where $j = 1, 2, 3, 4$, be the corresponding roots of (6-40), obtained by applying the transformation (6-35) to the primed roots. Let an electromagnetic wave in the moving magneto-plasma be represented by:

$$\overline{E}'^{(j)} = (\hat{x}E_{x'}^{(j)} + \hat{y}E_{y'}^{(j)})e^{i(\omega' t' - k_0' n_j' z')} \qquad (8\text{-}16)$$

$$\overline{H}'^{(j)} = (\hat{x}H_{x'}^{(j)} + \hat{y}H_{y'}^{(j)})e^{i(\omega' t' - k_0' n_j' z')} \qquad (8\text{-}17)$$

The field components in (8-16) satisfy:

$$E_{y'}^{(1)} = iE_{x'}^{(1)}; \quad E_{y'}^{(2)} = -iE_{x'}^{(2)} \tag{8-18}$$

$$E_{y'}^{(3)} = iE_{x'}^{(3)}; \quad E_{y'}^{(4)} = -iE_{x'}^{(4)} \tag{8-19}$$

which is a consequence of the electric field wave equation and the corresponding refractive index equation, satisfied by the characteristic waves of the magneto-plasma [Budden, 1961]. Using (8-16) - (8-17) in the Maxwell equation:

$$\nabla' \times \overline{E}'^{(j)} = -i\omega'\mu_o\overline{H}'^{(j)} \tag{8-20}$$

one obtains:

$$\eta_o H_{x'}^{(j)} = -n_j' E_{y'}^{(j)}; \quad \eta_o H_{y'}^{(j)} = n_j' E_{x'}^{(j)}, \quad j = 1, 2, 3, 4. \tag{8-21}$$

The electric fields of the waves in the plasma will transform in accordance with (2-13) as follows:

$$E_{x'}^{(j)} = \gamma(1 - n_j\beta_L)E_x^{(j)}; \quad E_{y'}^{(j)} = \gamma(1 - n_j\beta_L)E_y^{(j)} \tag{8-22}$$

The wave frequencies ω_j in the plasma will transform as in (6-27) - (6-28), and they are given in (8-59).

8.3 Reflection and Transmission Coefficients

The incident, reflected, and transmitted waves and the waves in the magneto-plasma slab must satisfy the tangential boundary conditions of the electromagnetic fields in the S' system [Landau and Lifshitz, 1966], as in (6-54) - (6-57). Substituting (8-1) - (8-6) and (8-16) - (8-17) in the tangential boundary conditions at z' = 0 in (6-54) - (6-57), one obtains:

$$\sum_{j=1}^{4} E_{x'}^{(j)} = E_{x'}^{I} + E_{x'}^{R} \tag{8-23}$$

$$\sum_{j=1}^{4} E_{y'}^{(j)} = E_{y'}^{R} \tag{8-24}$$

$$\eta_o \sum_{j=1}^{4} H_{x'}^{(j)} = \eta_o H_{x'}^{R} \tag{8-25}$$

$$\eta_o \sum_{j=1}^{4} H_{y'}^{(j)} = \eta_o H_{y'}^{I} + \eta_o H_{y'}^{R} \tag{8-26}$$

From the corresponding tangential boundary conditions at z' = d', one obtains:

$$\sum_{j=1}^{4} E_{x'}^{(j)} e^{-ik'_o n'_j d'} = E_{x'}^{T} e^{-ik'_o d'} \tag{8-27}$$

$$\sum_{j=1}^{4} E_{y'}^{(j)} e^{-ik'_o n'_j d'} = E_{y'}^{T} e^{-ik'_o d'} \tag{8-28}$$

$$\eta_o \sum_{j=1}^{4} H_{x'}^{(j)} e^{-ik'_o n'_j d'} = \eta_o H_{x'}^{T} e^{-ik'_o d'} \tag{8-29}$$

$$\eta_o \sum_{j=1}^{4} H_{y'}^{(j)} e^{-ik'_o n'_j d'} = \eta_o H_{y'}^{T} e^{-ik'_o d'} \tag{8-30}$$

The left-hand sides in the above equations (8-23) - (8-30) represent total fields on the plasma side of the boundaries, and the right-hand sides represent the total fields on the free space side of the boundaries.

Substituting (8-7), (8-18) - (8-19), (8-21) - (8-22), and (8-8) - (8-10) in (8-23) - (8-30), one obtains the following matrix equation for 8 unknowns in the laboratory system S in terms of the incident wave, where the common factor γ cancels throughout:

$$\left[\begin{array}{cccc}
(1-n_1\beta_L) & (1-n_2\beta_L) & (1-n_3\beta_L) & (1-n_4\beta_L) \\
i(1-n_1\beta_L) & -i(1-n_2\beta_L) & i(1-n_3\beta_L) & -i(1-n_4\beta_L) \\
i(n_1-\beta_L) & -i(n_2-\beta_L) & i(n_3-\beta_L) & -i(n_4-\beta_L) \\
(n_1-\beta_L) & (n_2-\beta_L) & (n_3-\beta_L) & (n_4-\beta_L) \\
\Lambda_1(1-n_1\beta_L) & \Lambda_2(1-n_2\beta_L) & \Lambda_3(1-n_3\beta_L) & \Lambda_4(1-n_4\beta_L) \\
i\Lambda_1(1-n_1\beta_L) & -i\Lambda_2(1-n_2\beta_L) & i\Lambda_3(1-n_3\beta_L) & -i\Lambda_4(1-n_4\beta_L) \\
-i\Lambda_1(n_1-\beta_L) & i\Lambda_2(n_2-\beta_L) & -i\Lambda_3(n_3-\beta_L) & i\Lambda_4(n_4-\beta_L) \\
\Lambda_1(n_1-\beta_L) & \Lambda_2(n_2-\beta_L) & \Lambda_3(n_3-\beta_L) & \Lambda_4(n_4-\beta_L)
\end{array}\right.$$

$$\begin{bmatrix} -(1+\beta_L) & 0 & 0 & 0 \\ 0 & -(1+\beta_L) & 0 & 0 \\ 0 & (1+\beta_L) & 0 & 0 \\ (1+\beta_L) & 0 & 0 & 0 \\ 0 & 0 & -(1-\beta_L) & 0 \\ 0 & 0 & 0 & -(1-\beta_L) \\ 0 & 0 & 0 & (1-\beta_L) \\ 0 & 0 & -(1-\beta_L) & 0 \end{bmatrix} \begin{bmatrix} E_x^{(1)} \\ E_x^{(2)} \\ E_x^{(3)} \\ E_x^{(4)} \\ E_x^R \\ E_y^R \\ E_x^T \\ E_y^T \end{bmatrix} = \begin{bmatrix} (1-\beta_L) \\ 0 \\ 0 \\ (1-\beta_L) \\ 0 \\ 0 \\ 0 \\ 0 \end{bmatrix} E_x^I \tag{8-31}$$

where we define from (6-36) and (8-11):

$$\Lambda_j = e^{ik_o'(1-n_j')d'} = e^{ik_o d'\sqrt{1-\beta_L^2}(1-n_j)/(1-n_j\beta_L)} \tag{8-32}$$

The algebraic expressions for the eight unknowns in (8-31) can be found after some algebraic manipulations; however, since the ultimate goal is to find the numerical results, (8-31) may as well be solved by a matrix inversion on a digital computer.

For the reflection and the transmission coefficients, only the last four unknowns need to be considered here. Let us define:

$$r_1 = E_x^R/E_x^I; \quad r_2 = E_y^R/E_x^I \tag{8-33}$$

$$t_1 = E_x^T/E_x^I; \quad t_2 = E_y^T/E_x^I \tag{8-34}$$

The power reflection and the power transmission coefficients may be defined as follows:

$$R = -\overline{n}\cdot\overline{S}_r/\overline{n}\cdot\overline{S}_i; \quad T = \overline{n}\cdot\overline{S}_t/\overline{n}\cdot\overline{S}_i \tag{8-35}$$

where we have:

$$\overline{S}_i = \frac{1}{2}(\overline{E}^I \times \overline{H}^{I*})$$

$$\overline{S}_r = \frac{1}{2}(\overline{E}^R \times \overline{H}^{R*})$$

$$\overline{S}_t = \frac{1}{2}(\overline{E}^T \times \overline{H}^{T*}),$$

and $\overline{n}$ is the unit vector normal to the upper interface. By substituting (8-33) - (8-34) in (8-35), one obtains:

$$R = |r_1|^2 + |r_2|^2; \quad T = |t_1|^2 + |t_2|^2. \tag{8-36}$$

In the following we shall present the numerical results for R and T versus β_L for certain particular cases.

8.4 Numerical Results and Discussion

Numerical results for the power reflection coefficient R and the power transmission coefficient T were obtained for $\Omega = 0.707(X = \omega_p^2/\omega^2 = 2.0)$, $k_o d' = \pi/4$, and $\Omega_H = 0., 0.5, 1.5$, where we define $\Omega = \omega/\omega_p$ and $\Omega_H = \omega_H/\omega_p$. Figure 14 shows the plot of the power reflection coefficient R versus $\beta_L = v_o/c$ for different values of the parameter Ω_H. The plot for $\Omega_H = 0$ agrees with the result obtained by Yeh [1966] (note the change in the sign for β_L) for the same case. The effect of the presence of the magnetostatic field is shown in the plots for $\Omega_H \neq 0$, and it is seen that the reflection coefficient R becomes more oscillatory. The absolute maximum attained by R increases with Ω_H for the present case. However, for a different Ω or $k_o d'$, this may not be the case. As $\beta_L \rightarrow -1$ for the present case, the reflection with the magneto-static field present is larger than it is for the case when it is not present. Comparing with the results of the semi-infinite plasma, it is found that the maximum reflection coefficient obtained there is greater than that for the present slab. This may be attributed to the larger kinetic energy associated with the semi-infinite moving plasma. It is further noted that $R < 1$ for $\beta_L > 0$. R may be larger than unity for $\beta_L < 0$; it reaches an absolute maximum in the range $-1. < \beta < 0$. Rapid oscillations in R are noted as $\beta_L \rightarrow -1$. This may be attributed to an increased number of resonance modes in the presence of the magneto-static field. Figure 15 shows the plot of the power transmission coefficient T versus $\beta_L = v_o/c$ for the same parameters as in figure 14. The plot for $\Omega_H = 0$ agrees with the result obtained by Yeh [1966]. The presence of the magneto-static field brings out some interesting results. Whereas, for the case of $\Omega_H = 0$, the transmission coefficient $T \rightarrow 0$ as $\beta_L \rightarrow 1$, we find that for the case of $\Omega_H \neq 0$, $T \rightarrow 1$ as $\beta_L \rightarrow 1$. This can be explained in terms of the refractive index observed in the rest system of the plasma. For the isotropic case at $\beta_L = 0.99$, $n'^2 = -397$, thus the wave is highly evanescent and presents a bad match with the free space where $n' = 1$. For the case of $\Omega_H = 0.5$ at $\beta_L = 0.99$, one obtains $n_1'^2 = 6.7$ and $n_2'^2 = -4.55$. One of the waves thus propagates without attenuation, while the other wave suffers attenuation, but not as large as in the case of the isotropic plasma. Most of the transmitted power is then carried by the propagating mode. It should be noted that an evanescent wave in the rest system of the plasma will appear to be a propagating wave with attenuation in the

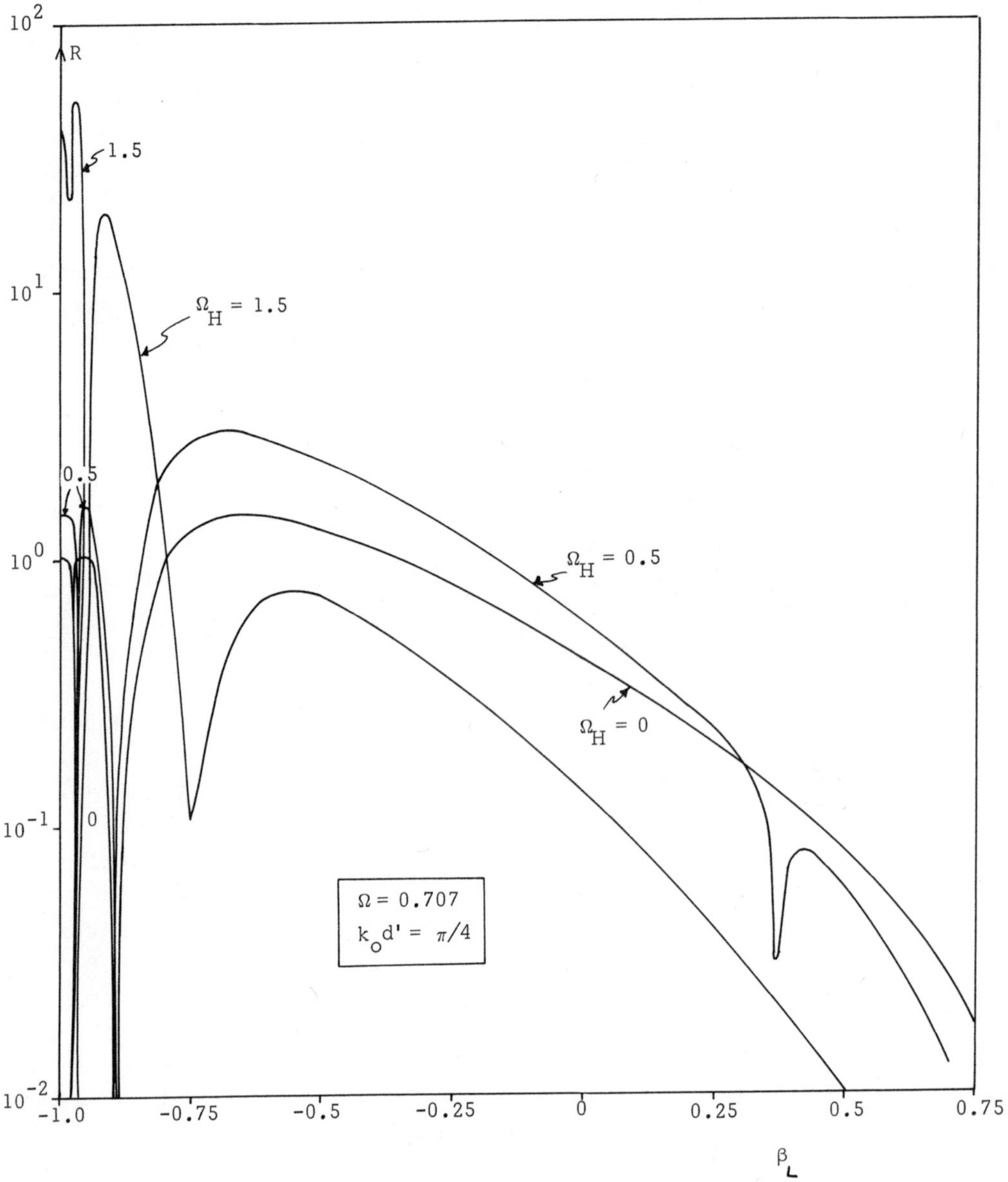

Figure 14. Reflection Coefficient vs. Normalized Velocity

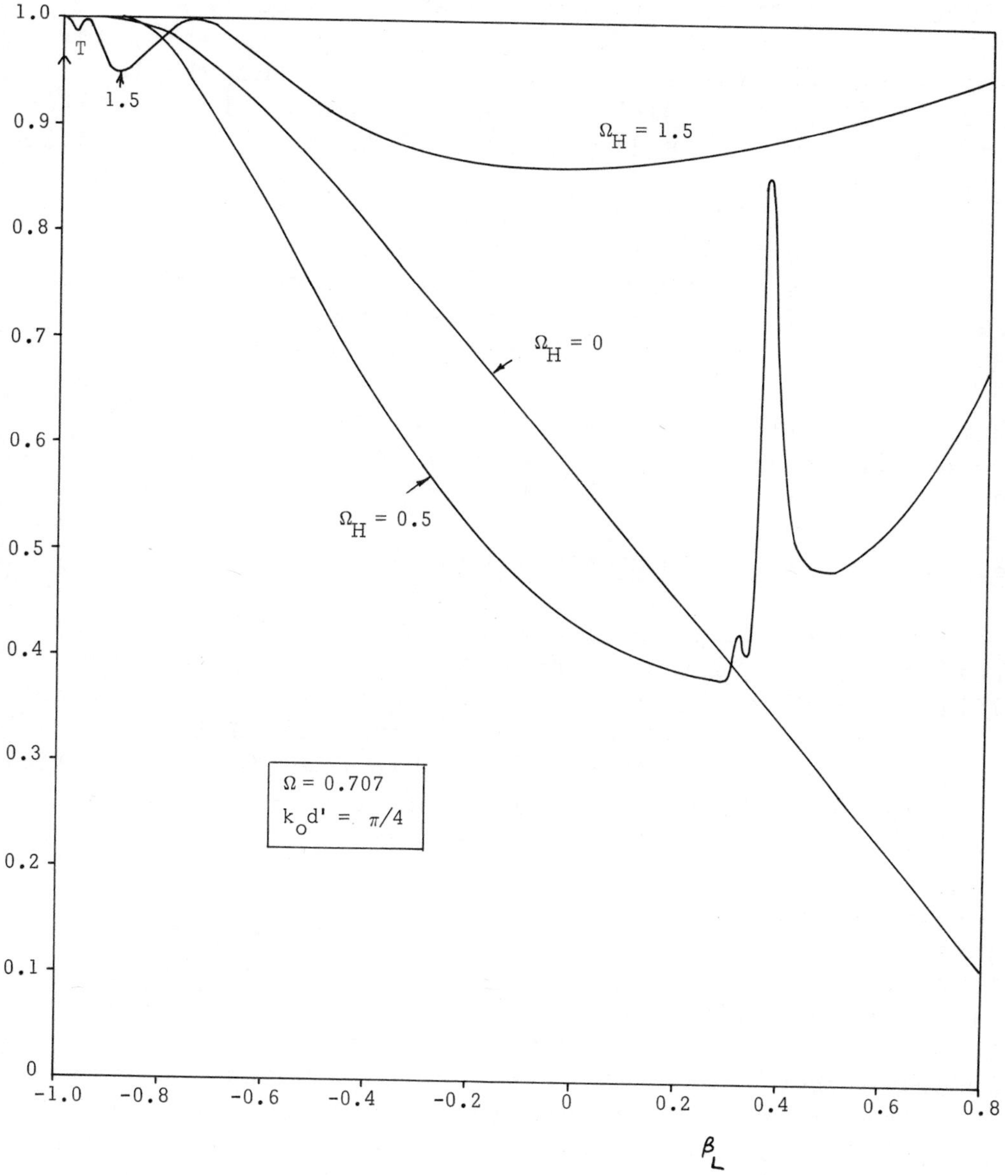

Figure 15. Transmission Coefficient vs. Normalized Velocity

laboratory system. The result for $\Omega_H = 1.5$ is rather interesting, since it is similar to the result for the moving dielectric slab as obtained by Yeh and Casey [1966]. The ripples in the transmission coefficient which were very small for $\Omega_H = 0.$, 0.5 as $\beta_L \rightarrow -1$, now become significant and are comparable to the ripples for the dielectric slab case. The transmission coefficient for $\Omega_H = 1.5$ is larger overall than the transmission coefficient for $\Omega_H = 0.5$. Thus an increase in the magneto-static field seems to make the plasma more transparent to the incident electromagnetic wave. It is to be noted that the transmission coefficient is less than unity for all β_L, unlike the reflection coefficient which could be larger than unity for $\beta_L < 0$.

The sum of the two coefficients $(R+T)$ is not equal to unity for any β different from zero. It is found that $(R+T) < 1$ for $\beta_L > 0$, $(R+T) > 1$ for $\beta_L < 0$, and, of course, $R+T = 1$ for $\beta_L = 0$.

8.5 Circularly Polarized Waves

In the present section, we will discuss the reflection and transmission coefficients, for a circularly polarized electromagnetic wave, normally incident on a plasma slab, moving along a longitudinally directed, uniform, static magnetic field. Since an incident circularly polarized wave excites an electromagnetic wave in the magneto-plasma of the same sense of circular polarization, the polarization of the reflected wave and the transmitted wave will also be circular of the same sense of polarization. The final result in this case is found to be in a closed form. It was shown in chapter six that both the relativistic and the non-relativistic solutions for the normal incidence on a moving, semi-infinite magneto-plasma, for the longitudinal case, give identical results. In the present section the case of circularly polarized waves will be solved by using the non-relativistic solution as in section 6.6, but identical results could be found, by using the relativistic solution as in section 8.3. The results in the present section will apply for both the non-relativistic and the relativistic solution.

Let a circularly polarized electromagnetic wave, normally incident on a plasma slab of width d, moving along a longitudinal static magnetic field, in the region $z \leqq v_0 t$, be of the form:

$$\overline{E}_{Ir} = E_r^I(\hat{x} - i\hat{y})e^{i(\omega_I t - k_I z)} \qquad (8\text{-}37)$$

$$\overline{H}_{Ir} = \frac{i}{\eta_0} E_r^I(\hat{x} - i\hat{y})e^{i(\omega_I t - k_I z)} \qquad (8\text{-}38)$$

$$\overline{E}_{I\ell} = E_\ell^I(\hat{x} + i\hat{y})e^{i(\omega_I t - k_I z)} \qquad (8\text{-}39)$$

$$\overline{H}_{I\ell} = \frac{-i}{\eta_0} E_\ell^I(\hat{x} + i\hat{y})e^{i(\omega_I t - k_I z)} \qquad (8\text{-}40)$$

where E_r^I and E_ℓ^I are constants, $\mathbf{k}_I = \omega_I/c$, $\eta_0 = \sqrt{\mu_0/\varepsilon_0}$, ω_I being the circular frequency of the incident electromagnetic wave in free space. The fields $\overline{E}_{Ir}$ and $\overline{H}_{Ir}$ in (8-37) - (8-38) represent the right-handed circularly polarized wave, which rotates in the clockwise direction, when one looks in the direction of propagation of the wave $\hat{z}$ [Budden, 1961]. The fields $\overline{E}_{I\ell}$ and $\overline{H}_{I\ell}$ in (8-39) - (8-40) represent the left-handed circularly polarized wave, which rotates in the counter-clockwise direction, when one looks in the direction of propagation of the wave $\hat{z}$ [Budden, 1961].

Similarly, the transmitted circularly polarized wave, beyond the plasma slab, in the region $z \geqq (v_0 t + d)$, will be of the form:

$$\overline{E}_{Tr} = E_r^T(\hat{x} - i\hat{y})e^{i(\omega_T t - k_T z)} \qquad (8\text{-}41)$$

$$\overline{H}_{Tr} = \frac{i}{\eta_0} E_r^T(\hat{x} - i\hat{y})e^{i(\omega_T t - k_T z)} \qquad (8\text{-}42)$$

$$\overline{E}_{T\ell} = E_\ell^T(\hat{x} + i\hat{y})e^{i(\omega_T t - k_T z)} \qquad (8\text{-}43)$$

$$\overline{H}_{T\ell} = -\frac{i}{\eta_0} E_\ell^T(\hat{x} + i\hat{y})e^{i(\omega_T t - k_T z)} \qquad (8\text{-}44)$$

where E_r^T, E_ℓ^T are constants and $k_T = \omega_T/c$, ω_T being the circular frequency of the transmitted wave. Equations (8-41) - (8-42) represent the right-handed circularly polarized transmitted wave, and (8-43) - (8-44) represent the left-handed circularly polarized transmitted wave.

The circularly polarized reflected wave in the region $z \leqq v_0 t$ will be of the form:

$$\overline{E}_{R\ell} = E_\ell^R(\hat{x} - i\hat{y})e^{i(\omega_R t + k_R z)} \qquad (8\text{-}45)$$

$$\bar{H}_{R\ell} = -\frac{i}{\eta_o} E_\ell^R (\hat{x} - i\hat{y}) e^{i(\omega_R t + k_R z)} \tag{8-46}$$

$$\bar{E}_{Rr} = E_r^R (\hat{x} + i\hat{y}) e^{i(\omega_R t + k_R z)} \tag{8-47}$$

$$\bar{H}_{Rr} = \frac{i}{\eta_o} E_r^R (\hat{x} + i\hat{y}) e^{i(\omega_R t + k_R z)} \tag{8-48}$$

where E_ℓ^R and E_r^R are constants and $k_R = \omega_R/c$, ω_R being the circular frequency of the reflected wave. Comparing (8-45) - (8-48) with (8-37) - (8-40), one finds that the right-handed circularly polarized incident wave corresponds to the left-handed circularly polarized reflected wave, and vice versa, since the two waves propagate in opposite directions, and the sense of the circular polarization is defined with respect to the direction of propagation [Budden, 1961].

The corresponding circularly polarized waves in the moving magneto-plasma slab $v_o t \leqq z \leqq (v_o t + d)$ will be of the form:

$$\bar{E}_1 = E^{(1)} (\hat{x} + i\hat{y}) e^{i(\omega_1 t - k_1 z)} \tag{8-49}$$

$$\bar{H}_1 = -i \frac{n_1}{\eta_o} E^{(1)} (\hat{x} + i\hat{y}) e^{i(\omega_1 t - k_1 z)} \tag{8-50}$$

$$\bar{E}_2 = E^{(2)} (\hat{x} - i\hat{y}) e^{i(\omega_2 t - k_2 z)} \tag{8-51}$$

$$\bar{H}_2 = i \frac{n_2}{\eta_o} E^{(2)} (\hat{x} - i\hat{y}) e^{i(\omega_2 t - k_2 z)} \tag{8-52}$$

$$\bar{E}_3 = E^{(3)} (\hat{x} + i\hat{y}) e^{i(\omega_3 t - k_3 z)} \tag{8-53}$$

$$\bar{H}_3 = -i \frac{n_3}{\eta_o} E^{(3)} (\hat{x} + i\hat{y}) e^{i(\omega_3 t - k_3 z)} \tag{8-54}$$

$$\bar{E}_4 = E^{(4)} (\hat{x} - i\hat{y}) e^{i(\omega_4 t - k_4 z)} \tag{8-55}$$

$$\bar{H}_4 = i \frac{n_4}{\eta_o} E^{(4)} (\hat{x} - i\hat{y}) e^{i(\omega_4 t - k_4 z)} \tag{8-56}$$

where $k_j = n_j \omega_j / c$, n_j being the corresponding refracting index solution of the double quadratic equation (6-20) or (6-40). We will have right-handed or left-handed circular polarization, depending on whether we take the solution $+|n_j|$ or $-|n_j|$, since it depends on the direction of propagation.

The boundary conditions (6-15) - (6-16) should apply at the moving plasma boundaries $z = v_o t$ and $z = (v_o t + d)$. Using (8-37) - (8-56) in (6-15) - (6-16), one has the following requirements for the exponential factors:

$$\omega_I t - k_I v_o t = \omega_R t + k_R v_o t = \omega_j t - k_j v_o t = \omega_T t - k_T v_o t \qquad (8\text{-}57)$$

Equation (8-57) may be rewritten in the form:

$$\omega_I (1 - \beta_L) = \omega_R (1 + \beta_L) = \omega_j (1 - n_j \beta_L) = \omega_T (1 - \beta_L) \qquad (8\text{-}58)$$

From (8-58) one obtains:

$$\omega_R = \omega_I \frac{1 - \beta_L}{1 + \beta_L} \; ; \; \omega_T = \omega_I ; \; \omega_j = \omega_I \frac{1 - \beta_L}{1 - n_j \beta_L} \qquad (8\text{-}59)$$

where (8-58) - (8-59) agree with the result for the relativistic case in (8-14), where $\omega \equiv \omega_I$.

Substituting (8-37) - (8-56) in the boundary condition (6-15) for the boundary $z = v_o t$, canceling the common exponential factor in accordance with (8-57), and separating between the $(\hat{x} + i\hat{y})$ and $(\hat{x} - i\hat{y})$ sense of polarization, one obtains:

$$(1 - n_1 \beta_L) E^{(1)} + (1 - n_3 \beta_L) E^{(3)} - (1 + \beta_L) E_r^R = (1 - \beta_L) E_\ell^I \qquad (8\text{-}60)$$

$$(1 - n_2 \beta_L) E^{(2)} + (1 - n_4 \beta_L) E^{(4)} - (1 + \beta_L) E_\ell^R = (1 - \beta_L) E_r^I \qquad (8\text{-}61)$$

Similarly, by substituting in the boundary condition (6-16), for the boundary $z = v_o t$, one obtains:

$$(n_1 - \beta_L) E^{(1)} + (n_3 - \beta_L) E^{(3)} + (1 + \beta_L) E_r^R = (1 - \beta_L) E_\ell^I \qquad (8\text{-}62)$$

$$(n_2 - \beta_L) E^{(2)} + (n_4 - \beta_L) E^{(4)} + (1 + \beta_L) E_\ell^R = (1 - \beta_L) E_r^I \qquad (8\text{-}63)$$

Substituting (8-37) - (8-56) in the boundary condition (6-15) for the boundary $z = (v_o t + d)$, one obtains:

$$\Lambda_1 (1 - n_1 \beta_L) E^{(1)} + \Lambda_3 (1 - n_3 \beta_L) E^{(3)} - (1 - \beta_L) E_\ell^T = 0 \qquad (8\text{-}64)$$

$$\Lambda_2 (1 - n_2 \beta_L) E^{(2)} + \Lambda_4 (1 - n_4 \beta_L) E^{(4)} - (1 - \beta_L) E_r^T = 0 \qquad (8\text{-}65)$$

Similarly, one obtains from the boundary condition (6-16) for the boundary $z = (v_o t + d)$:

$$\Lambda_1(n_1 - \beta_L)E^{(1)} + \Lambda_3(n_3 - \beta_L)E^{(3)} - (1-\beta_L)E_\ell^T = 0 \qquad (8\text{-}66)$$

$$\Lambda_2(n_2 - \beta_L)E^{(2)} + \Lambda_4(n_4 - \beta_L)E^{(4)} - (1-\beta_L)E_r^T = 0 \qquad (8\text{-}67)$$

where one defines $\Lambda_j = e^{i(k_T - k_j)d}$, where d is the width of the moving plasma slab in the laboratory stationary coordinate system. From the above definitions and (8-59), one has:

$$k_T = \frac{\omega_T}{c} = \frac{\omega_I}{c} = k_o;\; k_j = \frac{n_j\omega_j}{c} = k_o n_j \frac{1-\beta_L}{1-n_j\beta_L} \qquad (8\text{-}68)$$

From (8-68) one obtains:

$$(k_T - k_j)d = k_o d\left(1 - n_j \frac{1-\beta_L}{1-n_j\beta_L}\right) =$$

$$= k_o d \frac{1-n_j}{1-n_j\beta_L} = k_o d' \sqrt{1-\beta_L^2}\frac{1-n_j}{1-n_j\beta_L} \qquad (8\text{-}69)$$

where d' is the width of the non-moving plasma slab in the rest coordinate system, and d is the corresponding width of the moving plasma slab in the laboratory coordinate system, and in accordince with the Lorentz-Fitzgerald contraction [Pauli, 1958], one has:

$$d = d'/\gamma = d'\sqrt{1-\beta_L^2}.$$

From (8-69) we see that Λ_j is identical with the one given in (8-32).

Equations (8-60) - (8-63) agree, for the semi-infinite moving plasma, with (6-104) - (6-107). From (6-100) - (6-103) one has:

$$E_x^I = E_r^I + E_\ell^I;\quad iE_y^I = E_r^I - E_\ell^I \qquad (8\text{-}70)$$

$$E_x^T = E_r^T + E_\ell^T;\quad iE_y^T = E_r^T - E_\ell^T \qquad (8\text{-}71)$$

$$E_x^R = E_\ell^R + E_r^R;\quad iE_y^R = E_\ell^R - E_r^R \qquad (8\text{-}72)$$

Taking the sum and the difference of each pair of (8-60) - (8-67), and substituting (8-70) - (8-72), taking $E_y^I = 0$, one obtains the matrix relationship (8-31) found above for the relativistic case.

Equations (8-60) - (8-67) could be separated into two sets of four equations with four unknowns each. The first set will involve the unknown constants $E^{(1)}$,

$E^{(3)}$, E_r^R, E_ℓ^T in terms of E_ℓ^I, and the second set will involve the uhknown constants $E^{(2)}$, $E^{(4)}$, E_ℓ^R, and E_r^T in terms of E_r^I. Solving (8-60), (8-62), (8-64) and (8-66), one obtains:

$$\frac{E_r^R}{E_\ell^I} = r_{\ell r} = \frac{(1 - n_1)(1-n_3)(\Lambda_3 - \Lambda_1)}{(1+n_1)(1-n_3)\Lambda_3 - (1-n_1)(1+n_3)\Lambda_1} \tag{8-73}$$

$$\frac{E_\ell^T}{E_\ell^I} = t_{\ell\ell} = \frac{2(n_1-n_3)\Lambda_1\Lambda_3}{(1+n_1)(1-n_3)\Lambda_3 - (1-n_1)(1+n_3)\Lambda_1} \tag{8-74}$$

Similarly, solving (8-61), (8-63), (8-65), and (8-67), one obtains:

$$\frac{E_\ell^R}{E_r^I} = r_{r\ell} = \frac{(1-n_2)(1-n_4)(\Lambda_4 - \Lambda_2)}{(1 + n_2)(1-n_4)\Lambda_4 - (1-n_2)(1+n_4)\Lambda_2} \tag{8-75}$$

$$\frac{E_r^T}{E_r^I} = t_{rr} = \frac{2(n_2-n_4)\Lambda_2\Lambda_4}{(1+n_2)(1-n_4)\Lambda_4 - (1-n_2)(1+n_4)\Lambda_2} \tag{8-76}$$

where $r_{rr} = r_{\ell\ell} = t_{r\ell} = t_{\ell r} = 0$ in accordance with the discussion in section 6.6. While the solution of the circularly polarized incident wave case involves the inversion of a 4 x 4 matrix, the solution of the linearly polarized incident wave case in (8-31) involves the inversion of a 8 x 8 matrix. The solution of (8-31) of the linearly polarized incident wave could be found as a superposition of the solutions of two circularly polarized incident waves, in accordance with (8-70) - (8-72). The solutions (8-73) - (8-76) are similar to the ones obtained previously [Stratton, 1941] for the isotropic slab case.

The results (8-73) - (8-76) apply for the relativistic case as well, and may be derived from the relativistic solution, presented in sections 8.2 and 8.3. This may be accomplished by assuming that the incident plane electromagnetic wave has both the electric field components E_x^I and E_y^I in the laboratory system of coordinates S, and the corresponding components in the rest system of the moving plasma S'. By using the corresponding relativistic Lorentz transformations, one finds that for the present case, the left-hand side of the 8 x 8 matrix relationship (8-31) remains unaltered, while the right-hand side of (8-31) will have the following column matrix:

$$(1 - \beta_L)(E_x^I, E_y^I, E_y^I, E_x^I, 0,0,0,0) \qquad (8\text{-}77)$$

By assuming a circularly polarized incident electromagnetic wave $E_y^I = iE_x^I$ or $E_y^I = -iE_x^I$, and taking the sum and the difference of consecutive pairs, one may finally obtain (8-73) - (8-76) for the relativistic solution in identical form.

IX

OBLIQUE INCIDENCE ON SEMI-INFINITE TEMPERATE GENERAL MAGNETO-PLASMA MOVING THROUGH FREE SPACE

9.1 Introduction

In the present chapter a further generalization of chapter six is considered. The oblique electromagnetic wave is assumed to be incident in an arbitrary direction on a semi-infinite, temperate plasma, moving uniformly at an arbitrary direction with respect to a magnetostatic field. The power reflection coefficients for this case are derived by using the relativistic solution. The frequency, the wave vector, and the fields of the incident oblique electromagnetic wave are first transformed to the rest system of the plasma, where the boundary conditions are applied to the incident, reflected, and transmitted waves. The transmitted wave characteristics are given in terms of the plasma parameters as observed in the laboratory system by using the plasma parameter transformations. The reflected and the transmitted waves in the rest system of the plasma are then transformed to the laboratory system of coordinates, and the power reflection coefficients are found. Two different polarizations of the incident electromagnetic wave are considered: The parallel polarization and the perpendicular polarization. For the particular case of normal incidence, the results obtained are shown to agree with the previous results. Many other particular cases may be derived from the general relativistic results presented in the present chapter.

9.2 Formulation of the Problem

Let the plasma be at rest in a reference system S'(x',y',z',t'), moving uniformly with respect to a laboratory reference system S(x,y,z,t), such that their spatial axes are parallel. The system S' is assumed to be moving along the +z direction with a velocity

$\bar{v}_0(0, 0, v_0)$. The boundary of the semi-infinite plasma is assumed to be along the plane $z' = 0$, and the plasma is confined to the region $z' \leq 0$. A plane electromagnetic wave, with parallel or perpendicular polarization, has its wave vector in the x-z plane of incidence, and is incident from the right on the plasma boundary at an angle θ_i with the negative z axis. The primed quantities in the following analysis refer to the S' system, while the unprimed quantities refer to the S system. The geometry of the problem is shown in figure 16.

Let the incident wave in free space with its two possible polarizations, be represented in the two systems of coordinates, the rest system S' of the plasma and the laboratory system S.

For the parallel polarization case one has:

$$\bar{E}^I = (\hat{x}E_x^I + \hat{z}E_z^I)\psi_i ; \quad \bar{H}^I = -\hat{y}H_y^I \psi_i \tag{9-1}$$

$$\bar{E}'^I = (\hat{x}E_x'^I + \hat{z}E_z'^I)\psi_i' ; \quad \bar{H}'^I = -\hat{y}H_y'^I \psi_i' \tag{9-2}$$

where we define in general:

$$\psi_i = e^{i(\omega t - k_x x - k_z z)} ; \quad \psi_i' = e^{i(\omega' t' - k_x' x' - k_z' z')}$$

For the perpendicular polarization case, one has:

$$\bar{E}^I = \hat{y}E_y^I \psi_i ; \quad \bar{H}^I = (\hat{x}H_x^I + \hat{z}H_z^I)\psi_i \tag{9-3}$$

$$\bar{E}'^I = \hat{y}E_y'^I \psi_i' ; \quad \bar{H}'^I = (\hat{x}H_x'^I + \hat{z}H_z'^I)\psi_i' \tag{9-4}$$

By using the Maxwell equations and the Lorentz transformations for (9-1) - (9-2) from chapter two, one obtains for the parallel polarization case:

$$E_x^I = \eta_0 \cos\theta_i H_y^I ; \quad E_z^I = \eta_0 \sin\theta_i H_y^I \tag{9-5}$$

$$E_x'^I = \gamma\eta_0(\cos\theta_i + \beta)H_y^I ; \quad E_z'^I = E_z^I = \eta_0 \sin\theta_i H_y^I ;$$

$$H_y'^I = \gamma(1 + \beta\cos\theta_i)H_y^I \tag{9-6}$$

Similarly, by using them for (9-3) - (9-4), one obtains for the perpendicular polarization case:

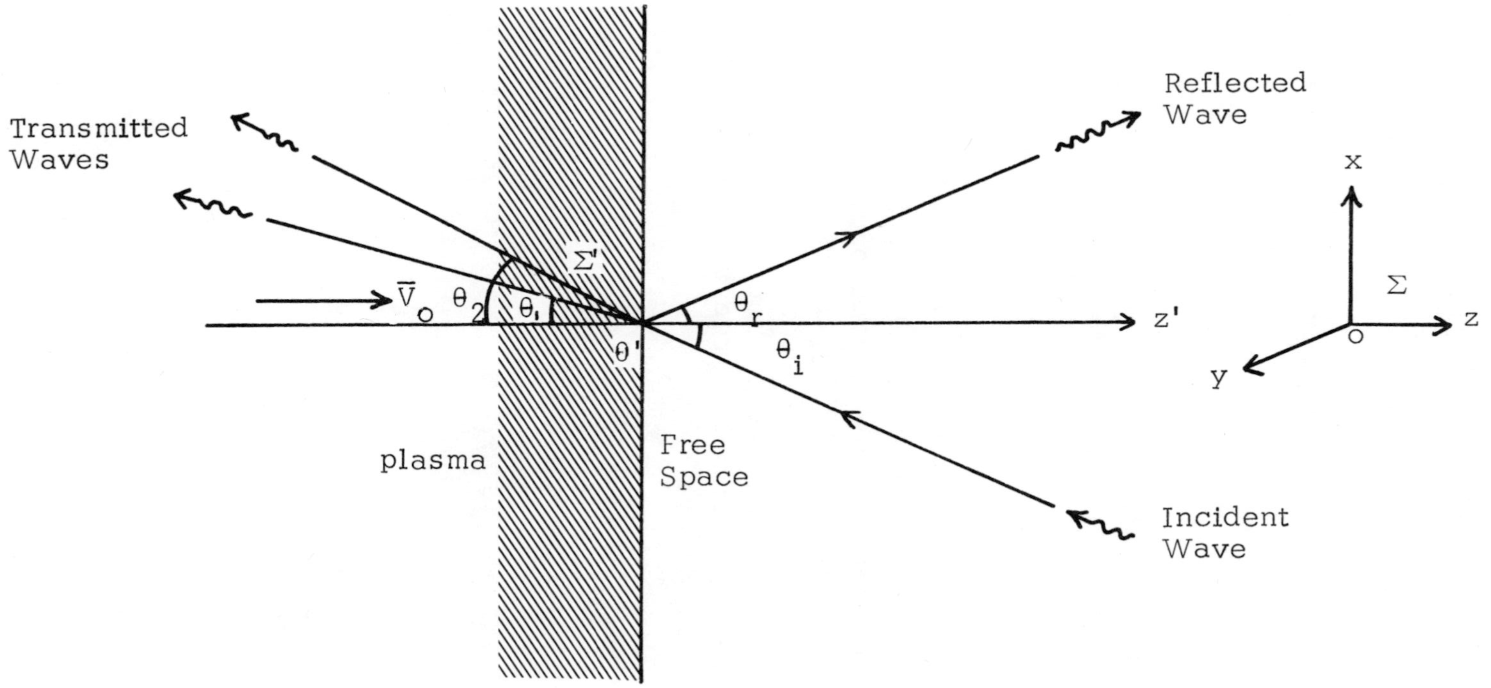

Figure 16. Oblique Incidence on Moving Semi-Infinite Magneto-Plasma

$$\eta_o H_x^I = \cos\theta_i E_y^I; \quad \eta_o H_z^I = \sin\theta_i E_y^I \tag{9-7}$$

$$E_y'^I = \gamma(1+\beta\cos\theta_i)E_y^I; \quad \eta_o H_x'^I = \gamma(\cos\theta_i + \beta)E_y^I;$$
$$\eta_o H_z'^I = \sin\theta_i E_y^I \tag{9-8}$$

For both cases the wave-frequency relations are:

$$k_x = k_i \sin\theta_i; \; k_z = -k_i \cos\theta_i \tag{9-9}$$

$$k_x' = k_x = k_i \sin\theta_i; \; k_z' = \gamma(k_z - \frac{\omega_i v_o}{c^2}),$$
$$\omega' = \gamma\omega_i(1+\beta\cos\theta_i) \tag{9-10}$$

where we define $k_i = \omega_i\sqrt{\mu_o\,\epsilon_o} = \omega_i/c$, ω_i being the incident wave circular frequency, $\eta_o = \sqrt{\mu_o/\epsilon_o}$, $\gamma = (1-\beta^2)^{-1/2}$, and $\beta = \beta_L = v_o/c$.

Let us consider now the reflected wave. From th studies of reflection from a stationary, semi-infinite magneto-plasma [Budden, 1961], it is known that the reflected wave, in general, does not have the same polarization as the incident wave. Let the reflected wave in free space be represented in the two systems of coordinates S and S' by:

$$\overline{E}^R = (\hat{x}E_x^R + \hat{y}E_y^R + \hat{z}E_z^R)\psi_r; \; \overline{H}^R = (\hat{x}H_x^R + \hat{y}H_y^R + \hat{z}H_z^R)\psi_r \tag{9-11}$$

$$\overline{E}'^R = (\hat{x}E_x'^R + \hat{y}E_y'^R + \hat{z}E_z'^R)\psi_r'; \; \overline{H}'^R = (\hat{x}H_x'^R + \hat{y}H_y'^R + \hat{z}H_z'^R)\psi_r' \tag{9-12}$$

where we define in general:

$$\psi_r = e^{i(\omega_r t - k_{rx}x + k_{rz}z)}; \; \psi_r' = e^{i(\omega' t' - k_x' x' + k_z' z')}.$$

The use of the Maxwell equations and the application of the Lorentz transformations for (9-11) - (9-12) from chapter two give:

$$\eta_o H_x^R = -\cos\theta_r E_y^R; \quad \eta_o H_z^R = \sin\theta_r E_y^R;$$
$$E_x^R = \eta_o \cos\theta_r H_y^R; \quad E_z^R = -\eta_o \sin\theta_r H_y^R \tag{9-13}$$

$$E_x'^R = \gamma(\cos\theta_r - \beta)\eta_o H_y^R;\quad E_y'^R = \gamma(1-\beta\cos\theta_r)E_y^R;$$

$$\eta_o H_x'^R = -\gamma(\cos\theta_r - \beta)E_y^R;\quad \eta_o H_y'^R = \gamma(1-\beta\cos\theta_r)\eta_o H_y^R;$$

$$E_z'^R = -\eta_o \sin\theta_r H_y^R;\quad \eta_o H_z'^R = \sin\theta_r E_y^R \tag{9-14}$$

$$k_{rx} = k_r \sin\theta_r = k_x' = k_i \sin\theta_i;$$

$$k_{rz} = -k_r\cos\theta_r = -\gamma^2 k_i[(1+\beta^2)\cos\theta_i + 2\beta];$$

$$k_r = \omega_r/c;\quad \omega_r = \gamma^2\omega_i[(1+\beta^2) + 2\beta\cos\theta_i];$$

$$\sin\theta_r = \sin\theta_i/\gamma^2[(1+\beta^2)+2\beta\cos\theta_i];$$

$$\cos\theta_r = \frac{1}{\beta}\{1-(1+\beta\cos\theta_i)/\gamma^2[(1+\beta^2)+2\beta\cos\theta_i]\} \tag{9-15}$$

where $\sin^2\theta_r + \cos^2\theta_r = 1$.

Let us now consider the waves transmitted in the plasma. The fields in the plasma in the S' system satisfy the Maxwell equations:

$$\nabla' \times \overline{E}' = -i\omega'\mu_o\overline{H}' \tag{9-16}$$

$$\nabla' \times \overline{H}' = i\omega'\epsilon_o\overline{E}' + \overline{\overline{\sigma}}'\cdot\overline{E}' \tag{9-17}$$

where $\overline{\overline{\sigma}}'$ is the conductivity tensor [Brandstatter, 1963]. Let us take the factor $e^{-i(k_x'x'+k'q'z')}$, where $k' = \omega'/c$, to be the space variation of a characteristic wave propagating in the plasma. Substituting this factor in (9-16) - (9-17), one may obtain three linear, homogeneous equations with three unknowns E_x', E_y', E_z', after eliminating $\overline{H}'$. The equation may be written in the following matrix form [Budden, 1961; Unz, 1965b]:

$$\begin{bmatrix} A_{11}' & A_{12}' & A_{13}' \\ A_{21}' & A_{22}' & A_{23}' \\ A_{31}' & A_{32}' & A_{33}' \end{bmatrix} \begin{bmatrix} E_x' \\ E_y' \\ E_z' \end{bmatrix} = \begin{bmatrix} 0 \\ 0 \\ 0 \end{bmatrix} \tag{9-18}$$

where we define:

$$A_{11}' = q'^2 - 1 + X'(U'^2 - Y_1'^2)/D'$$

$$A'_{12} = X'(iU'Y'_3 - Y'_1Y'_2)/D'$$

$$A'_{13} = -S'_iq' + X'(-iU'Y'_2 - Y'_1Y'_3)/D'$$

$$A'_{21} = X'(-iU'Y'_3 - Y'_1Y'_2)/D'$$

$$A'_{22} = S'^2_i + q'^2 - 1 + X'(U'^2 - Y'^2_2)/D'$$

$$A'_{23} = X'(iU'Y'_1 - Y'_2Y'_3)/D'$$

$$A'_{31} = -S'_iq' + X'(iU'Y'_2 - Y'_1Y'_3)/D'$$

$$A'_{32} = X'(-iU'Y'_1 - Y'_2Y'_3)/D'$$

$$A'_{33} = S'^2_i - 1 + X'(U'^2 - Y'^2_3)/D'$$

$$D' = U'(U'^2 - Y'^2)$$

$$U' = 1 - iZ', \quad Z' = \nu'/\omega'$$

$$\overline{Y}' = (Y'_1, Y'_2, Y'_3) = e\overline{B}'_o/m'\omega' = \overline{\omega}'_H/\omega'$$

$$S'_i = \sin\theta'_i = \sin\theta_i/\gamma(1 + \beta\cos\theta_i)$$

The condition for a non-trivial solution of (9-18) requires that the determinant of the matrix $\overline{\overline{A}}'$ be made equal to zero. This gives an algebraic equation of the fourth order in q', thus representing four waves. Two of those waves exist in the plasma region $z' < 0$ in figure 16, as transmitted waves traveling away from the boundary. The Booker quartic algebraic equation thus found is as follows [Budden, 1961; Unz, 1965b]:

$$\alpha'_4q'^4 + \alpha'_3q'^3 + \alpha'_2q'^2 + \alpha'_1q' + \alpha'_o = 0 \qquad (9\text{-}19)$$

where

$$\alpha'_4 = U'^2(U' - X') - U'Y'^2 + X'Y'^2_3$$

$$\alpha'_3 = 2X'S'_iY'_1Y'_3$$

$$\alpha_2' = 2(C'^2U'-X')[Y'^2-(U'-X')U']+X'(Y'^2+S_i'^2Y_1'^2-C'^2Y_3'^2)$$

$$\alpha_1' = -2C'^2S_i'X'Y_1'Y_3'$$

$$\alpha_0' = (C'^2U'-X')[(C'^2U'-X')(U'-X')-C'^2Y'^2]-C'^2S_i'^2X'Y_1'^2$$

where $Y'^2 = Y_1'^2 + Y_2'^2 + Y_3'^2$, $C'^2 = 1 - S_i'^2$, and $Y_1' = Y_x'$, $Y_2' = Y_y'$, $Y_3' = Y_z'$.

Let the four roots of (9-19) be denoted by q_j', j = 1, 2, 3, 4. Let the subscripts 1 and 2 represent the waves excited in $z' \leq 0$ in the semi-infinite plasma.

The equation (9-19) is in terms of the quantities in the rest system S'. The coefficients α_j' can be transformed to the laboratory system S, by using the plasma parameters and the wave vector-frequency transformations. The plasma parameters transformations are given in section 2.4 as follows:

$$\omega_p' = \omega_p; \quad \nu' = \gamma\nu; \quad \overline{\omega}_H' = (\omega_{H1}', \omega_{H2}', \omega_{H3}') = (\gamma^2\omega_{H1}, \gamma^2\omega_{H2}, \gamma\omega_{H3}). \tag{9-20}$$

Using $\omega' = \gamma\omega_i(1 + \beta\cos\theta_i)$ and (9-20), the plasma parameters X', Y', Z', and U' may be written in terms of X, Y, Z, and U. Substituting these new plasma parameters in (9-19), the resulting Booker quartic equation may then be solved for q_j' in terms of the plasma parameters in the laboratory system S. Having obtained q_j' in the above manner, one may then obtain q_j as follows, by using Appendix B:

$$q_j = (q_j' + \beta)/(1 + q_j'\beta), \quad q_j' = (q_j - \beta)/1 - q_j\beta) \tag{9-21}$$

The transmitted wave has the following properties in the laboratory system S:

$$k_{tx}^j = k_t^j \sin\theta_t^j; \quad k_{tz}^j = k_t^j q_j \tag{9-22}$$

$$k_t^j = \omega_t^j/c; \quad \omega_t^j = \omega_i(1 + \beta\cos\theta_i)/(1 - q_j\beta) \tag{9-23}$$

$$\sin\theta_t^j = \sin\theta_i(1 - q_j\beta)/(1 + \beta\cos\theta_i) \tag{9-24}$$

$$n_j^2 = q_j^2 + [\sin\theta_i(1 - q_j\beta)/(1 + \beta\cos\theta_i)]^2 \tag{9-25}$$

where n_j is the refractive index of the jth plasma wave. The angle θ_t^j is the complex angle made by the transmitted wave with the negative z-axis in the plasma.

The equation (9-18) may be used to express the y' and z' components of the electric field in terms of the x' component. Using the last two of the three linear homogeneous equations represented by (9-18), one obtains:

$$E_y'^j/E_x'^j = (A_{23}'^j\, A_{31}'^j - A_{21}'^j\, A_{33}'^j)/(A_{22}'^j\, A_{33}'^j - A_{23}'^j\, A_{32}'^j) = p_j' \tag{9-26}$$

$$E_z'^j/E_x'^j = (A_{21}'^j\, A_{32}'^j - A_{22}'^j\, A_{31}'^j)/(A_{22}'^j\, A_{33}'^j - A_{23}'^j\, A_{32}'^j) = u_j' \tag{9-27}$$

Let A_{ij}^k be the elements of the matrix $\bar{\bar{A}}^k$ after the transformations (9-10), (9-20), (9-21), and (9-22) - (9-25) have been substituted to express the elements of the matrix $\bar{\bar{A}}'^k$ in terms of the laboratory parameters. Let p_j and u_j also represent p_j' and u_j' after the substitution in the above manner.

The Lorentz transformations of the fields associated with the transmitted waves in the plasma will be given by:

$$E_x'^j = \gamma(1 - \beta q_j)\ell_j E_x^j; \quad E_y'^j = p_j' E_x'^j = \gamma(1 - \beta q_j) p_j \ell_j E_x^j \tag{9-28}$$

$$\eta_0 H_x'^j = \gamma(\beta - q_j) p_j \ell_j E_x^j; \quad \eta_0 H_y'^j = \gamma w_j E_x^j \tag{9-29}$$

where we define:

$$w_j = (q_j - \beta) - \gamma n_j \sin\theta_t^j (1 - \beta q_j) u_j \ell_j$$

$$\ell_j = 1/(1 - \beta n_j \sin\theta_t^j u_j).$$

9.3 Reflection and Transmission Coefficients

The boundary conditions in S' require the continuity of the tangential components of the electric and magnetic fields at z' = 0 [Landau and Lifshitz, 1966]. Substituting for the incident, reflected, and transmitted waves in the boundary conditions at z' = 0 in (6-54) - (6-57), and using the Lorentz field transformations given above, one obtains for the parallel polarization case, after canceling out the common factor γ:

$$\begin{bmatrix} (1-\beta q_1)\ell_1 & (1-\beta q_2)\ell_2 & -(\cos\theta_r-\beta) & 0 \\ (1-\beta q_1)\ell_1 p_1 & (1-\beta q_2)\ell_2 p_2 & 0 & -(1-\beta\cos\theta_r) \\ (\beta-q_1)\ell_1 p_1 & (\beta-q_2)\ell_2 p_2 & 0 & (\cos\theta_r-\beta) \\ w_1 & w_2 & -(1-\beta\cos\theta_r) & 0 \end{bmatrix} \begin{bmatrix} E_x^1 \\ E_x^2 \\ \eta_o H_y^R \\ E_y^R \end{bmatrix} = \begin{bmatrix} (\cos\theta_i+\beta) \\ 0 \\ 0 \\ (1+\beta\cos\theta_i) \end{bmatrix} \eta_o H_y^I \tag{9-30}$$

and similarly for the perpendicular polarization case, finds:

$$\begin{bmatrix} (1-\beta q_1)\ell_1 & (1-\beta q_2)\ell_2 & -(\cos\theta_r-\beta) & 0 \\ (1-\beta q_1)\ell_1 p_1 & (1-\beta q_2)\ell_2 p_2 & 0 & -(1-\beta\cos\theta_r) \\ (\beta-q_1)\ell_1 p_1 & (\beta-q_2)\ell_2 p_2 & 0 & (\cos\theta_r-\beta) \\ w_1 & w_2 & -(1-\beta\cos\theta_r) & 0 \end{bmatrix} \begin{bmatrix} E_x^1 \\ E_x^2 \\ \eta_o H_y^R \\ E_y^R \end{bmatrix} = \begin{bmatrix} 0 \\ (1+\beta\cos\theta_i) \\ (\cos\theta_i+\beta) \\ 0 \end{bmatrix} E_y^I \tag{9-31}$$

For the particular case of normal incidence $\theta_i = 0$, $\theta_t^j = \theta_r = 0$, taking the magneto-static field along the z direction, equation (9-30) agrees with the results of chapter six. For this same particular case, (9-31) gives the wave amplitudes when the normally incident wave has the component E_y^I, instead of E_x^I.

For the particular case of normal incidence, taking the magneto-static field transverse to the direction of motion, along the $\hat{y}$ direction, one has $E_x'(1) = E_z'(1) = 0$ and $p_2 = 0$. Equation (9-30) can then be modified into the form:

$$\begin{bmatrix} 0 & (1-\beta q_2) & -(1-\beta) & 0 \\ (1-\beta q_1) & 0 & 0 & -(1-\beta) \\ (\beta-q_1) & 0 & 0 & (1-\beta) \\ 0 & (q_2-\beta) & -(1-\beta) & 0 \end{bmatrix} \begin{bmatrix} E_y^{(1)} \\ E_x^{(2)} \\ \eta_o H_y^R \\ E_y^R \end{bmatrix} = \begin{bmatrix} 1+\beta \\ 0 \\ 0 \\ 1+\beta \end{bmatrix} \eta_o H_y^I \tag{9-32}$$

where equation (9-32) agrees with the corresponding result of (7-88) of chapter seven, for the particular case $\kappa = 1$.

Equations (9-30) and (9-31) can be readily solved for the four unknowns by matrix inversion.

Having found all the unknowns in this manner, the power reflection coefficient may then be defined as:

$$R = -\hat{n} \cdot \bar{E}_R \times \bar{H}_R^* / \hat{n} \cdot \bar{E}_I \times \bar{H}_I^* \tag{9-33}$$

where $\hat{n}$ is the unit vector normal to the boundary ($\hat{z}$ in this case), and the corresponding complex Poynting vectors are given by:

$$\bar{E}_R \times \bar{H}_R^* = \frac{1}{\eta_o} (|\eta_o H_y^R|^2 + |E_y^R|^2)(\hat{x} \sin \theta_r + \hat{z} \cos \theta_r)$$

$$\hat{n} \cdot \bar{E}_R \times \bar{H}_R^* = \frac{\cos \theta_r}{\eta_o} \cdot (|\eta_o H_y^R|^2 + |E_y^R|^2)$$

for parallel polarization one has:

$$\hat{n} \cdot \bar{E}_I \times \bar{H}_I^* = \frac{\cos \theta_i}{\eta_o} \, |\eta_o H_y^I|^2$$

for perpendicular polarization one has:

$$\hat{n} \cdot \bar{E}_I \times \bar{H}_I^* = \frac{\cos \theta_i}{\eta_o} \, |E_y^I|^2$$

Using (9-33) and the above, one finds for parallel polarization:

$$R_1 = \frac{\cos \theta_r}{\cos \theta_i} (|r_{11}|^2 + |r_{12}|^2) \tag{9-34}$$

and for perpendicular polarization:

$$R_2 = \frac{\cos \theta_r}{\cos \theta_i} (|r_{21}|^2 + |r_{22}|^2) \tag{9-35}$$

where we define:

$$r_{11} = \eta_o H_y^R / \eta_o H_y^I ; \quad r_{12} = E_y^R / \eta_o H_y^I$$

$$r_{21} = \eta_o H_y^R / E_y^I ; \quad r_{22} = E_y^R / E_y^I$$

It should be noted that the transmission coefficient may not be defined in this manner, since the transmitted wave frequency and the wave vector as observed in the system S, are not always real.

X

OBLIQUE INCIDENCE ON WARM GENERAL MAGNETO-PLASMA SLAB MOVING THROUGH A DIELECTRIC

10.1 Introduction

In the present chapter, the formal solution of the general problem of oblique incidence of an electromagnetic wave on a warm general magneto-plasma slab, moving through a dielectric, is presented. The solutions obtained in chapters six through nine may be found as particular cases of the present solution. The particular cases of the general refractive index equation, in the rest frame S' of the plasma, are shown to agree with the results of the previous chapters. The plasma slab is assumed to be moving uniformly in the +z direction, which is also normal to the boundaries, and the static magnetic field is assumed to be arbitrarily directed. The stationary infinite dielectric medium may not be a real dielectric, but may be represented by a dielectric in the sense described in chapter seven. A linearly polarized electromagnetic wave is assumed to be obliquely incident at an angle with respect to the boundary of the slab. In the rest system S' of the plasma slab, it is found that six waves are excited. The field components of the reflected and transmitted waves are given by a matrix relation in terms of the incident wave fields. The power reflection and the power transmission coefficients are defined as in chapter eight. The linearized, single fluid, continuum theory of plasma dynamics and the Minkowski expressions for moving media are used in order to find the characteristic equation in the S' system.

10.2 Formulation of the Problem

Let a warm plasma slab bounded by the planes z' = 0 and z' = -d' in the rest system S'(x',y',z',t') be moving uniformly along the +z direction with respect to a laboratory system S (x,y,z,t), which has (Fig. 17) its spatial axes x, y, z parallel to x', y', z', respectively. The dielectric medium of permittivity $\epsilon = \epsilon_0 \kappa$ is stationary in the unprimed laboratory system S. An electromagnetic wave traveling in the dielectric region z' > 0 is assumed to be obliquely incident on the plane z' = 0, making an angle θ_i with the +z axis. The solutions for the reflected and transmitted waves are sought. The procedure to find the solutions is the same as before; first the incident, reflected, and transmitted waves are transformed to the S' rest system, where the boundary conditions are applied at the two boundaries of the plasma slab, taking into account the six waves in the plasma. Next the wave and the plasma parameter transformations are applied, and the electromagnetic fields of the reflected, the transmitted, and the six waves in the plasma are obtained in terms of the incident wave fields, as a set of ten linear inhomogeneous equations. Only two field components of the reflected and transmitted waves and one component of the wave in the plasma are involved in this equation. The rest of the fields of the waves can be determined by the use of Maxwell equations.

Let an oblique electromagnetic wave represented by:

$$\overline{E}_I = (\hat{x}E_x^{\ I} + \hat{y}E_y^{\ I} + \hat{z}E_z^{\ I})\,\psi_i\,;\ \overline{H}_I = (\hat{x}H_x^{\ I} + \hat{y}H_y^{\ I} + \hat{z}H_z^{\ I})\,\psi_i \quad (10\text{-}1)$$

$$\overline{E}_I' = (\hat{x}E_x'^{\ I} + \hat{y}E_y'^{\ I} + \hat{z}E_z'^{\ I})\,\psi_i'\,;\ \overline{H}_I' = (\hat{x}H_x'^{\ I} + \hat{y}H_y'^{\ I} + \hat{z}H_z'^{\ I})\,\psi_i' \quad (10\text{-}2)$$

in the S and the S' systems, respectively, be incident in the x-z plane of incidence, at an angle θ_i with the z axis, where we define:

$$\psi_i = e^{i(\omega_i t - k_x x - k_z z)}\ ;\ \psi_i' = e^{i(\omega' t' - k_x' x' - k_z' z')}$$

$$k_x = \omega_i M \sin\theta_i / c;\ k_z = -\omega_i M \cos\theta_i / c$$

where $M = \sqrt{\kappa}$ and $\kappa = \epsilon/\epsilon_0$ is the dielectric constant of the stationary dielectric medium.

Using the Maxwell equations:

$$\nabla \times \overline{E} = -i\omega\mu_0 \overline{H};\ \nabla \times \overline{H} = i\omega\epsilon\overline{E}, \quad (10\text{-}3)$$

the components $E_x^{\ I}$, $H_x^{\ I}$, $E_z^{\ I}$, and $H_z^{\ I}$ can be written in terms

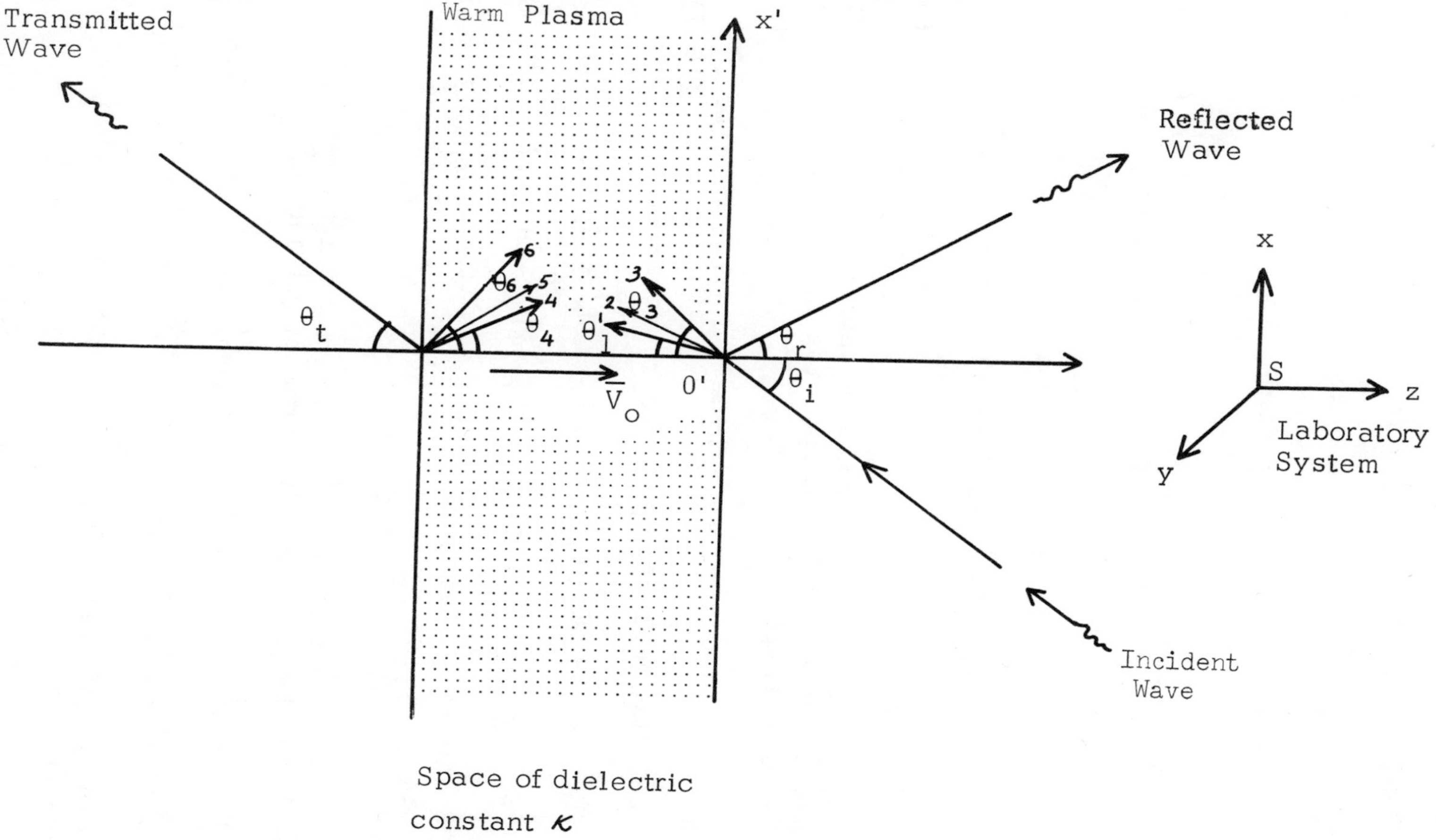

Figure 17. Oblique Incidence on a Plasma Slab Moving Through a Dielectric

of E_y^I and H_y^I as follows:

$$E_x^I = -\eta \cos \theta_i H_y^I; \quad \eta H_x^I = \cos \theta_i E_y^I \tag{10-4}$$

$$E_z^I = -\eta \sin \theta_i H_y^I; \quad \eta H_z^I = \sin \theta_i E_y^I \tag{10-5}$$

where $\eta = \sqrt{\mu_0/\varepsilon}$.

The primed field quantities appearing in (10-2) are given in terms of E_y^I and H_y^I, by using the Lorentz transformations in (2-13) and (2-16):

$$E_x'^I = -\gamma\eta(\cos \theta_i + M\beta)H_y^I; \quad \eta H_x'^I = \gamma(\cos \theta_i + M\beta)E_y^I;$$

$$E_y'^I = \gamma(1 + M\beta \cos \theta_i)E_y^I; \quad \eta H_y'^I = \gamma\eta(1 + M\beta \cos \theta_i)H_y^I;$$

$$E_z'^I = E_z^I; \quad H_z'^I = H_z^I \tag{10-6}$$

The frequency ω' and the wave-vector components k_x', k_z' are given by the corresponding Lorentz transformations (2-25) - (2-26):

$$\omega' = \gamma\omega_i(1+M\beta \cos \theta_i); \quad k_x' = k_x; \quad k_z' = -\gamma\omega_i(M \cos \theta_i + \beta)/c \tag{10-7}$$

Let the reflected wave be represented by:

$$\overline{E}_R = (\hat{x}E_x^R + \hat{y}E_y^R + \hat{z}E_z^R)\psi_r; \quad \overline{H}_R = (\hat{x}H_x^R + \hat{y}H_y^R + \hat{z}H_z^R)\psi_r \tag{10-8}$$

$$\overline{E}_R' = (\hat{x}E_x'^R + \hat{y}E_y'^R + \hat{z}E_z'^R)\psi_r'; \quad \overline{H}_R' = (\hat{x}H_x'^R + \hat{y}H_y'^R + \hat{z}H_z'^R)\psi_r' \tag{10-9}$$

in the S and S' systems, respectively, where we define:

$$\psi_r = e^{i(\omega_r t - k_{xr}x + k_{zr}z)}; \quad \psi_r' = e^{i(\omega' t' - k_x' x' + k_z' z')}$$

As before, the E_x^R, H_x^R, E_z^R, and H_z^R components in (10-8) can be obtained in terms of E_y^R and H_y^R:

$$E_x^R = \eta \cos \theta_r H_y^R; \quad \eta H_x^R = -\cos \theta_r E_y^R \tag{10-10}$$

$$E_z^R = -\eta \sin \theta_r H_y^R; \quad \eta H_z^R = \sin \theta_r E_y^R \tag{10-11}$$

where $\sin \theta_r = \sin \theta_i / \gamma^2 [(1 + \beta^2) + 2M\beta \cos \theta_i]$, which is a consequence of:

$$k_x' = k_x(=\omega_i M \sin \theta_i/c) = k_{xr}(=\omega_r M \sin \theta_r/c) \tag{10-12}$$

$$\omega_r = \gamma^2[(1+\beta^2) + 2M\beta\cos\theta_i]\omega_i;\ k_{zr} = -\omega_r M\cos\theta_r/c.$$

The primed field quantities in (10-9) are given by:

$$E_x'^R = \gamma\eta(\cos\theta_r - M\beta)H_y^R;\quad \eta H_x'^R = -\gamma(\cos\theta_r - M\beta)E_y^R$$

$$E_y'^R = \gamma(1 - M\beta\cos\theta_r)E_y^R;\quad \eta H_y'^R = \gamma\eta(1 - M\beta\cos\theta_r)H_y^R \quad (10\text{-}13)$$

The z-components will not be needed while applying boundary conditions.

Similar considerations apply to the transmitted wave in the free space region $z' < -d'$:

$$\overline{E}_T = (\hat{x}E_x^T + \hat{y}E_y^T + \hat{z}E_z^T)\psi_t;\ \overline{H}_T = (\hat{x}H_x^T + \hat{y}H_y^T + \hat{z}H_z^T)\psi_t \quad (10\text{-}14)$$

$$\overline{E}_T' = (\hat{x}E_x'^T + \hat{y}E_y'^T + \hat{z}E_z'^T)\psi_t';\ \overline{H}_T' = (\hat{x}H_x'^T + \hat{y}H_y'^T + \hat{z}H_z'^T)\psi_t' \quad (10\text{-}15)$$

in S and S' systems, respectively, where we define:

$$\psi_t = e^{i(\omega_t t - k_{xt}x - k_{zt}z)};\quad \psi_t' = e^{i(\omega' t' - k_{xt}'x' - k_{zt}'z')},$$

and we have:

$$\omega_t = \omega_i;\ k_{xt} = k_x;\ k_{zt} = k_z = -\omega_i M\cos\theta_i/c. \quad (10\text{-}16)$$

For (10-14) we obtain:

$$E_x^T = -\eta\cos\theta_i H_y^T;\quad \eta H_x^T = \cos\theta_i E_y^T \quad (10\text{-}17)$$

$$E_z^T = -\eta\sin\theta_i H_y^T;\quad \eta H_z^T = \sin\theta_i E_y^T \quad (10\text{-}18)$$

and we also obtain:

$$\begin{aligned} E_x'^T &= -\gamma\eta(\cos\theta_i + M\beta)H_y^T;\quad \eta H_x'^T = \gamma(\cos\theta_i + M\beta)E_y^T \\ E_y'^T &= \gamma(1 + M\beta\cos\theta_i)E_y^T;\quad \eta H_y'^T = \gamma\eta(1 + M\beta\cos\theta_i)H_y^T \end{aligned} \quad (10\text{-}19)$$

Let us now consider the waves in the warm plasma slab. Using the linearized, single fluid, continuum theory of plasma dynamics, the fields in the plasma in the S' system should satisfy [Oster, 1960]:

$$\nabla' \times \overline{E}' = -i\omega'\overline{B}' \quad (10\text{-}20)$$

$$\nabla' \times \overline{H}' = i\omega' \overline{D}' - eN_o' \overline{u}' \tag{10-21}$$

$$m' N_o' (\dot{\overline{u}}' + \nu' \overline{u}') = -\nabla' p' - eN_o' \overline{u}' - eN_o' \overline{u}' \times \overline{B}_o' \tag{10-22}$$

$$\nabla' p' = \alpha K T_o' \nabla' N' \tag{10-23}$$

$$\nabla' \cdot \overline{D}' = -eN' \tag{10-24}$$

where the quantities with the subscript o are the stationary values in S', and the non-subscripted ones are the perturbed quantities of small signal theory; N' is the perturbed electron density, $\overline{u}'$ is the perturbed electron velocity, p' is the perturbed pressure, T_o' is the uniform temperature of the plasma, K is the Boltzmann constant, α is the ratio of the electron gas specific heat at constant pressure to its specific heat at constant volume, $\overline{B}_o'$ is the externally impressed static magnetic field, and D' and $\overline{B}'$ are given by the Minkowski expressions (2-52) and (2-53) for the present case:

$$\overline{D}' = \epsilon \overline{\overline{\alpha}} \cdot \overline{E}' - \overline{\overline{\xi}} \cdot \overline{H}' \tag{10-25}$$

$$\overline{B}' = \mu_o \overline{\overline{\alpha}} \cdot \overline{H}' + \overline{\overline{\xi}} \cdot \overline{E}' \tag{10-26}$$

where we define:

$$\overline{\overline{\alpha}} = \begin{bmatrix} a & 0 & 0 \\ 0 & a & 0 \\ 0 & 0 & 1 \end{bmatrix} ; \quad \overline{\overline{\xi}} = \begin{bmatrix} 0 & -\xi & 0 \\ \xi & 0 & 0 \\ 0 & 0 & 0 \end{bmatrix} ;$$

$$a = (1 - \beta^2)/(1 - \kappa\beta^2); \quad \xi = (\kappa - 1)\beta/(1 - \kappa\beta^2)c$$

Equations (10-25) - (10-26) can be written in alternative vector form (7-4) - (7-5). It is then possible, at least in principle, to write a vector equation for $\overline{E}'$ by eliminating all the other variables from (10-20) - (10-26). Eliminating p' and N' from (10-22) - (10-24), one obtains:

$$eN_o' \overline{u}' = -\frac{i}{\omega'} \overline{\overline{M}}' \cdot \left[\frac{\alpha K T_o'}{m'} \nabla' (\nabla' \cdot \overline{D}') - \frac{e^2 N_o'}{m'} \overline{E}' \right] \tag{10-27}$$

where we define:

$$\overline{\overline{M}}' = \frac{1}{U'(U'^2 - Y_1'^2 - Y_2'^2 - Y_3'^2)} \begin{bmatrix} U'^2 - Y_1'^2 & -iU'Y_3' - Y_1'Y_2' & iU'Y_2' - Y_1'Y_3' \\ iU'Y_3' - Y_1'Y_2' & U'^2 - Y_2'^2 & -iU'Y_1' - Y_2'Y_3' \\ -iU'Y_2' - Y_1'Y_3' & iU'Y_1' - Y_2'Y_3' & U'^2 - Y_3'^2 \end{bmatrix}$$

Substituting (10-27) in (10-21), the corresponding Maxwell equation becomes:

$$\nabla' \times \overline{H}' = i\omega'\overline{D}' + \frac{i}{\omega'}\overline{\overline{M}}' \cdot \left[\frac{\alpha K T_o'}{m'} \nabla'(\nabla' \cdot \overline{D}') - \frac{e^2 N_o'}{m'}\overline{E}'\right] \qquad (10\text{-}28)$$

The Maxwell equations (10-20) and (10-28) reduce to (7-1) - (7-2) for temperate ($T_o' \simeq 0$) magneto-plasma moving through a dielectric, and to (9-16) - (9-17) for temperate magneto-plasma moving through free space.

Let the space variation of the waves in the plasma be $e^{-i(k_x'x' + k'q'z')}$, where $k' = \omega'/c$. An algebraic equation for q' can be found by substituting for the space variation and for $\overline{D}'$ and $\overline{B}'$ from (10-25) - (10-26) in (10-20) and (10-28) and by using the condition for non-trivial solution of the six homogeneous linear equations in six unknowns $\overline{E}'$ and $\overline{H}'$. The set of the six equations is found to be in the form:

$$(\overline{\overline{Q}} + i\omega'\overline{\overline{\xi}}) \cdot \overline{E}' + i\omega'\mu_o\overline{\overline{\alpha}} \cdot \overline{H}' = 0 \qquad (10\text{-}29)$$

$$\left[i\omega'\epsilon(\overline{\overline{I}} + \frac{c^2\delta'}{\omega'^2}\overline{\overline{M}}' \cdot \overline{\overline{\theta}}) \cdot \overline{\overline{\alpha}} - \frac{i}{\omega'}\frac{e^2N_o'}{m'}\overline{\overline{M}}'\right] \cdot \overline{E}' -$$

$$- \left[\overline{\overline{Q}} + i\omega'(\overline{\overline{I}} + \frac{c\,\delta'}{\omega'^2}\overline{\overline{M}}' \cdot \overline{\overline{\theta}})\right] \cdot \overline{\overline{\xi}} \cdot \overline{H}' = 0 \qquad (10\text{-}30)$$

where we define:

$$\overline{\overline{Q}} = \begin{bmatrix} 0 & ik'q' & 0 \\ -ik'q' & 0 & ik_x' \\ 0 & -ik_x' & 0 \end{bmatrix} ; \quad \overline{\overline{\theta}} = \begin{bmatrix} -k_x'^2 & 0 & -qk_x'k' \\ 0 & 0 & 0 \\ -qk_x'k' & 0 & -q'k'^2 \end{bmatrix}$$

and have $\delta' = a'^2/c^2$; $a'^2 = \alpha K T_o'/m'$, a' being the acoustic velocity in the electron gas.

Equations (10-29) - (10-30) represent a set of six homogeneous linear equations in six unknowns E_x', E_y',

E'_z, H'_x, H'_y, and H'_z. For a non-trivial solution the determinant of their coefficient matrix must be equal to zero. This gives an algebraic equation of the sixth order in q', thus representing six waves in the plasma slab. Let the six values of q' be represented as q'_j, $j = 1,\ldots,6$. The plasma parameter transformations of section 2.4 may be used to obtain the results of q'_j in terms of the plasma parameters observed in the laboratory system. Having obtained q'_j in the above manner, one may then obtain q_j by using the following relation, proved in Appendix B:

$$q_j = (q'_j + \beta)/(1 + q'_j\beta);\ q'_j = (q_j - \beta)/(1 - q_j\beta) \qquad (10\text{-}31)$$

Let each of the six waves in the plasma in the rest system S' be represented by:

$$\overline{E}'^j = (\hat{x}E'^j_x + \hat{y}E'^j_y + \hat{z}E'^j_z)\,\psi'_j;\ \overline{H}'^j = (\hat{x}H'^j_x + \hat{y}H'^j_y + \hat{z}H'^j_z)\,\psi'_j,\ j=1,\ldots 6 \qquad (10\text{-}32)$$

where we define:

$$\psi'_j = e^{i(\omega't' - k'_x x' - k'_o q'_j z')}.$$

Substituting q'_j for the jth wave in (10-29) - (10-30), one can solve for E'^j_y, E'^j_z, H'^j_x, H'^j_y, and H'^j_z in terms of E'^j_x by a matrix inversion applied to any five out of the six equations. Let us define:

$$E'^j_y/E'^j_x = p'_j;\ E'^j_z/E'^j_x = v'_j;$$

$$\eta H'^j_x/E'^j_x = w'_j;\ \eta H'^j_y/E'^j_x = h'_j;\ H'^j_z/E'^j_x = \ell'_j \qquad (10\text{-}33)$$

where it is assumed in (10-33) that the plasma parameter transformations are used in the coefficient matrix before inverting it. Similarly, by substituting (10-25) and (10-33) in (10-27) and using the parameter transformations, we define:

$$\beta'_j = u'^j_z/E'^j_x \qquad (10\text{-}34)$$

10.3 Reflection and Transmission Coefficients

We shall now substitute the above results in the boundary conditions at the planes z' = 0 and z' = -d'.

The boundary conditions in S' are the continuity of tangential electric and magnetic fields [Landau and Lifshitz, 1966], and the vanishing of the normal component of the velocity field $\bar{u}'$ at the boundaries, $\bar{u}' \cdot \hat{n} = 0$, where $\hat{n}$ is the normal to the boundary. The last boundary condition, $\bar{u}' \cdot \hat{n} = 0$, arises from the assumption that the plasma is confined to the region $-d' < z' < 0$, and the boundaries are assumed to be rigid in some sense [Unz, 1967b].

Substituting the incident, reflected and transmitted waves in free space and the six waves in the plasma slab, in the five boundary conditions at $z' = 0$, and in the five boundary conditions at $z' = -d'$, and defining $\Lambda_j' = e^{i(k'q_j' - k_z')d'}$, one obtains, for all x' and t', the fo lowing 10×10 matrix relationship:

$$
\left[\begin{matrix}
1 & 1 & 1 & 1 & 1 & 1 & 0 & -\gamma(\cos\theta_r - M\beta) \\
p_1' & p_2' & p_3' & p_4' & p_5' & p_6' & -\gamma(1-M\beta\cos\theta_r) & 0 \\
w_1' & w_2' & w_3' & w_4' & w_5' & w_6' & \gamma(\cos\theta_r - M\beta) & 0 \\
h_1' & h_2' & h_3' & h_4' & h_5' & h_6' & 0 & -\gamma(1-M\beta\cos\theta_r) \\
\beta_1' & \beta_2' & \beta_3' & \beta_4' & \beta_5' & \beta_6' & 0 & 0 \\
\Lambda_1' & \Lambda_2' & \Lambda_3' & \Lambda_4' & \Lambda_5' & \Lambda_6' & 0 & 0 \\
p_1'\Lambda_1' & p_2'\Lambda_2' & p_3'\Lambda_3' & p_4'\Lambda_4' & p_5'\Lambda_5' & p_6'\Lambda_6' & 0 & 0 \\
w_1'\Lambda_1' & w_2'\Lambda_2' & w_3'\Lambda_3' & w_4'\Lambda_4' & w_5'\Lambda_5' & w_6'\Lambda_6' & 0 & 0 \\
h_1'\Lambda_1' & h_2'\Lambda_2' & h_3'\Lambda_3' & h_4'\Lambda_4' & h_5'\Lambda_5' & h_6'\Lambda_6' & 0 & 0 \\
\beta_1'\Lambda_1' & \beta_2'\Lambda_2' & \beta_3'\Lambda_3' & \beta_4'\Lambda_4' & \beta_5'\Lambda_5' & \beta_6'\Lambda_6' & 0 & 0
\end{matrix}\right.
$$

$$
\left.\begin{array}{cc}
\cdots\; 0 & 0 \\
0 & 0 \\
0 & 0 \\
0 & 0 \\
0 & 0 \\
0 & \gamma(\cos\theta_i+M\beta) \\
-\gamma(1+M\beta\cos\theta_i) & 0 \\
-\gamma(\cos\theta_i+M\beta) & 0 \\
0 & -\gamma(1+M\beta\cos\theta_i) \\
0 & 0
\end{array}\right]
\begin{bmatrix}
E_x'^{1} \\ E_x'^{2} \\ E_x'^{3} \\ E_x'^{4} \\ E_x'^{5} \\ E_x'^{6} \\ E_y^{R} \\ \eta H_y^{R} \\ E_y^{T} \\ \eta H_y^{T}
\end{bmatrix}
=
\begin{bmatrix}
-\gamma\eta(\cos\theta_i+M\beta)H_y^{I} \\
\gamma(1+M\beta\cos\theta_i)E_y^{I} \\
\gamma(\cos\theta_i+M\beta)E_y^{I} \\
\gamma\eta(1+M\beta\cos\theta_i)H_y^{I} \\
0 \\ 0 \\ 0 \\ 0 \\ 0 \\ 0
\end{bmatrix}
\qquad (10\text{-}35)
$$

Equation (10-35) can be solved for E_y^R, ηH_y^R, E_y^T, and ηH_y^T in terms of the incident wave fields, by inverting the coefficient matrix on a digital computer.

The power reflection and the power transmission coefficients may then be defined as:

$$R = -\hat{n} \cdot \overline{E}_R \times \overline{H}_R^* / \hat{n} \cdot \overline{E}_I \times \overline{H}_I^* \tag{10-36}$$

$$T = \hat{n} \cdot \overline{E}_T \times \overline{H}_T^* / \hat{n} \cdot \overline{E}_I \times \overline{H}_I^* \tag{10-37}$$

where $\hat{n}$ is normal to the boundary, the star * represents the complex conjugate, and the cross products on the right are the complex Poynting vectors for the reflected, incident, and transmitted waves. Explicit expressions for R and T can be found similarly to those given at the end of the chapter nine.

The formal solution obtained above is very general. Numerical results for R and T for this general case, or any particular cases, can be found on a digital computer.

XI

RADIATION IN MOVING TEMPERATE MAGNETO-PLASMA

11.1 Introduction

Radiation from an oscillating dipole, in the presence of a moving medium, has received much attention in recent years. The earliest work dates back to Frank [1943], who analyzed the problem only in the rest frame of the medium. Unz [1963] presented the formal solution for the radiation fields in the presence of a moving, Tellegen, anisotropic medium. He formulated the problem in the rest frame of the source and used Chu's formulation of moving media [Fano, Chu, and Adler, 1960]. Lee and Papas [1964] solved the problem for a simple medium, with permittivity ϵ' and permeability μ' in the rest frame of the medium. They also formulated the problem in the rest frame of the source, making use of a four-vector potential differential equation. The problem of dipole radiation in a moving medium was solved independently by Tai [1965b,c], who used dyadic Green's functions for a moving isotropic medium. The potential equations for moving media, for relativistic velocities of the medium, are known to be derivable by two methods [Papas, 1965]. A third alternative derivation for non-relativistic velocities only was given by Chawla and Unz [1966c], and it may be found in Appendix A. Lee and Papas [1965] further presented a generalization of their previous work to moving isotropic plasmas, thus taking into account the dispersive nature of the moving medium. Here again the use of potentials was made, and the medium was assumed to have a dielectric constant $\kappa' = (1-\omega_p^2/\omega'^2)$ in its rest frame, but an approach through the electric field was described in their appendix.

It is well understood that the problems in electromagnetic theory can be looked upon from two different viewpoints [Brandstatter, 1963]:

a) the macroscopic phenomenological theory of Maxwell,

b) the exhaustive microscopic theory of Lorentz.

In the Maxwellian viewpoint, one describes the medium, such as plasma, to be a dielectric, characterized by an effective dielectric constant, or a dyadic, when the plasma is magnetically biased. For a stationary, isotropic plasma the relations between $\bar{D}$ and $\bar{E}$, and $\bar{B}$ and $\bar{H}$ are simple. The relations become rather complicate if the medium is anisotropic, such as magneto-plasma. The Minkowski theory of moving isotropic media was extended by Tai [1965a] to include anisotropic effects for non-relativistic velocities only. Extensions for relativistic velocities were presented by Chen and Cheng [1966] and by Lee and Lo [1966], who formulated the problem of radiation in a plasma, moving along a static magnetic field in the $\hat{z}$-direction. Certain difficulties in the integration were pointed out, when the above formulation was specialized to an infinite static magnetic field, in which case the dielectric tensor is diagonalized. The integrals were, however, evaluated in closed form, for moving, uniaxial crystal media. The problem here too, was formulated in the rest system of the source. A brief account of their formulation is given in section 11.2.

In the Lorentzian viewpoint [Brandstatter, 1963], one dispenses with the concept of a medium, and considers only the ensemble of negative and positive charges, which actually constitute the medium, in free space. A lucid description of this concept may be found in the reference cited. Using this concept, the problem of radiation from a source, immersed in a plasma, moving at an arbitrary angle with respect to a static magnetic field, was formulated in the rest frame of the source by Chawla and Unz [1966a], and is the subject of section 11.3.

11.2 Polarization Current Model (PCM)

In the present section, a brief review of the formulation by Lee and Lo [1966] is presented. The plasma in its rest frame is characterized by a dielectric tensor, thus involving a notion of polarization. For anisotropic media, characterized by permittivity $\bar{\bar{\epsilon}}'$ and permeability $\bar{\bar{\mu}}'$, the constitutive relations (2-52) -

(2-53) of the Minkowski theory may be written [Lee and Lo, 1966] in the form:

$$\bar{D} = \bar{\bar{\epsilon}} \cdot \bar{E} + \bar{\bar{\xi}} \cdot \bar{H} \tag{11-1}$$

$$\bar{B} = \bar{\bar{\eta}} \cdot \bar{E} + \bar{\bar{\mu}} \cdot \bar{H} \tag{11-2}$$

where $\bar{\bar{\epsilon}}$, $\bar{\bar{\xi}}$, $\bar{\bar{\eta}}$, and $\bar{\bar{\mu}}$ are found to be:

$$\bar{\bar{\epsilon}} = \epsilon_0\mu_0(c^2\bar{\bar{W}}\cdot\bar{\bar{\epsilon}}'\cdot\bar{\bar{W}} - c^2\bar{\bar{W}}\,\bar{\bar{\epsilon}}'\,\bar{\bar{W}}^{-1}\bar{\bar{V}}\,\bar{\bar{\mu}}'\bar{\bar{L}}\,\bar{\bar{\epsilon}}'\bar{\bar{W}}^{-1} + \bar{\bar{V}}\,\bar{\bar{L}}\,\bar{\bar{V}}\,\bar{\bar{\epsilon}}'\bar{\bar{W}}^{-1}) \tag{11-3}$$

$$\bar{\bar{\xi}} = \epsilon_0\mu_0(c^2\bar{\bar{W}}\,\bar{\bar{\epsilon}}'\,\bar{\bar{W}}^{-1}\bar{\bar{V}}\,\bar{\bar{\mu}}'\,\bar{\bar{L}} - \bar{\bar{V}}\,\bar{\bar{L}}) \tag{11-4}$$

$$\bar{\bar{\eta}} = \epsilon_0\mu_0(\bar{\bar{V}}\,\bar{\bar{W}}^{-1} - \bar{\bar{V}}\,\bar{\bar{W}}^{-1}\bar{\bar{V}}\,\bar{\bar{\mu}}'\bar{\bar{L}}\,\bar{\bar{V}}\,\bar{\bar{\epsilon}}'\bar{\bar{W}}^{-1} - c^2\bar{\bar{W}}\,\bar{\bar{\mu}}'\bar{\bar{L}}\,\bar{\bar{V}}\,\bar{\bar{\epsilon}}'\bar{\bar{W}}^{-1}) \tag{11-5}$$

$$\bar{\bar{\mu}} = \epsilon_0\mu_0(\bar{\bar{V}}\,\bar{\bar{W}}^{-1}\bar{\bar{V}}\,\bar{\bar{\mu}}'\bar{\bar{L}} + c^2\bar{\bar{W}}\,\bar{\bar{\mu}}'\,\bar{\bar{L}}) \tag{11-6}$$

where $\bar{\bar{V}} = \begin{bmatrix} 0 & -v_{oz} & v_{oy} \\ v_{oz} & 0 & -v_{ox} \\ -v_{oy} & v_{ox} & 0 \end{bmatrix}$, $\bar{\bar{W}} = \bar{\bar{I}} + \frac{1-\gamma}{\gamma v_o^2}\,\bar{v}_o\bar{v}_o^T$

and we define:

$\bar{\bar{I}}$ = identity matrix

$\gamma = (1-\beta^2)^{-1/2}$, $\beta = v_o/c$

$\bar{v}_o^T$ = the transpose of the column vector $\bar{v}_o$

$\bar{\bar{L}} = (\bar{\bar{W}} + \bar{\bar{V}}\,\bar{\bar{\epsilon}}'\,\bar{\bar{W}}^{-1}\bar{\bar{V}}\,\bar{\bar{\mu}}')^{-1}$

For a temperate magneto-plasma, moving along the static magnetic field in the $\hat{z}$ direction, one has in the rest system S' of the plasma:

$$\bar{\bar{\epsilon}}' = \epsilon_0 \begin{bmatrix} \epsilon_1' & i\epsilon_3' & 0 \\ -i\epsilon_3' & \epsilon_1' & 0 \\ 0 & 0 & \epsilon_2' \end{bmatrix}, \quad \bar{\bar{\mu}}' = \mu_0 \begin{bmatrix} 1 & 0 & 0 \\ 0 & 1 & 0 \\ 0 & 0 & 1 \end{bmatrix} \tag{11-7}$$

where $\epsilon_1' = 1 - X'/(1-Y'^2)$, $\epsilon_2' = 1 - X'$, $\epsilon_3' = +X'Y'/(1-Y'^2)$,

$X' = \omega_p'^2/\omega'^2$, and $Y' = \omega_H'/\omega'$; ω_p' and ω_H' are the plasma and cyclotron frequencies, respectively, in the rest system S', and ω' is the wave frequency in S' system, where harmonic time variation $e^{i\omega' t'}$ is assumed.

Substituting (11-7) in (11-3) - (11-6), one obtains:

$$\bar{\bar{\epsilon}} = \epsilon_0 \begin{bmatrix} M_1 & M_3 & 0 \\ -M_3 & M_1 & 0 \\ 0 & 0 & \epsilon_2' \end{bmatrix}; \quad \bar{\bar{\xi}} = \frac{\beta^2}{v_0} \begin{bmatrix} M_3 & -N_3 & 0 \\ N_3 & M_3 & 0 \\ 0 & 0 & 0 \end{bmatrix}; \tag{11-8}$$

$$\bar{\bar{\mu}} = \mu_0 \begin{bmatrix} Q_1 & \beta^2 M_3 & 0 \\ -\beta^2 M_3 & Q_1 & 0 \\ 0 & 0 & 1 \end{bmatrix}; \quad \bar{\bar{\eta}} = -\bar{\bar{\xi}} \tag{11-9}$$

where

$$M_1 = [\epsilon_1'(1 - \epsilon_1'\beta^2) + \beta^2\epsilon_3'^2]/Q_2$$

$$M_3 = i\,\epsilon_3'/Q_2, \; Q_1 = (1 - \epsilon_1'\beta^2)/Q_2$$

$$N_3 = [(\epsilon_1' - 1)(1 - \epsilon_1'\beta^2) + \beta^2\epsilon_3'^2]\gamma^2/Q_2$$

$$Q_2 = [(1 - \epsilon_1'\beta^2)^2 - (\epsilon_3'\beta^2)^2]\gamma^2.$$

The transformations of the plasma frequency and the cyclotron frequency in the S' system to those in the S system were given in section 2.4.

The Maxwell equations for the laboratory system S may be written as:

$$\nabla \times \bar{E} = -i\omega\bar{B} - \bar{M} \tag{11-10}$$

$$\nabla \times \bar{H} = i\omega\bar{D} + \bar{J} \tag{11-11}$$

where $\bar{M}$ and $\bar{J}$ are the source magnetic and electric current densities, respectively, and a harmonic time variation $e^{+i\omega t}$ is assumed.

The formal solution of the system of Maxwell equations (11-10) - (11-11) for $\bar{E}$ and $\bar{H}$ is then obtained by taking their Fourier transforms, and substituting for $\overline{B}$ and $\overline{D}$ the expressions (11-1) and (11-2), where $\overline{\overline{\epsilon}}$, $\overline{\overline{\xi}}$, $\overline{\overline{\mu}}$, and $\overline{\overline{\eta}}$ are defined in (11-8) - (11-9). A set of linear, inhomogeneous equations for the transformed quantity $\overline{E}(\overline{k})$ or $\overline{H}(\overline{k})$ is first obtained, by eliminations in the transformed equations. The linear inhomogeneous equation is then solved by matrix inversion. $\overline{E}(\overline{r})$ or $\overline{H}(\overline{r})$ is obtained by the inverse Fourier transforms. The details may be found in the original paper [Lee and Lo, 1966]. The difficulty in evaluating the inverse Fourier transform arises from the fact that the frequency ω' in the parameters described in (11-8) - (11-9) is the Doppler-shifted frequency

$$\omega' = \gamma(\omega - \overline{k} \cdot \overline{v}) \tag{11-12}$$

which appears in the integrand of the inverse Fourier transform integral, thus making the integral extremely difficult to be evaluated in a closed form.

11.3 Convection Current Model (CCM)

In the present section, the problem of radiation from sources, immersed in a plasma, moving at an arbitrary angle with respect to the static magnetic field, is formulated using the Lorentzian viewpoint. It will be found that the formulation presented here is less cumbersome. The Fourier transform method is used in presenting the formal solution, although no claim is made regarding the possible evaluation of the inverse Fourier integrals.

Let us assume, without a loss of generality, that the plasma as a whole is moving past the laboratory frame with a velocity v_o, in the $\hat{z}$-direction, where $\beta = \beta_L = v_0/c < 1$. The coordinate axes of the rest frame S' of the plasma are parallel to those of S. The static magnetic field is assumed to be arbitrarily directed.

Let the electric and the magnetic current sources $\overline{J}^s$ and $\overline{M}^s$ be located in the laboratory frame S. The plasma may be characterized in the S' frame by the following [Brandstatter, 1963]:

$$\overline{D}' = \epsilon_o \overline{E}' \tag{11-13}$$

$$\overline{B}' = \mu_o \overline{H}' \tag{11-14}$$

$$\overline{J}' = \overline{\overline{\sigma}}' \cdot \overline{E}' \tag{11-15}$$

where the conductivity tensor $\overline{\overline{\sigma}}'$ is defined [Ratcliffe, 1959; Budden, 1961; Brandstatter, 1963] in the following form:

$$\overline{\overline{\sigma}}' = - \frac{iN'e^2}{m'\omega'} \left[(1 - i \frac{\nu'}{\omega'}) \overline{\overline{I}} + i \overline{\overline{\Omega}}' \right]^{-1} \tag{11-16}$$

the gyro-tensor $\overline{\overline{\Omega}}'$ is defined as:

$$\overline{\overline{\Omega}}' = \frac{1}{\omega'} \begin{bmatrix} 0 & -\omega'_{H3} & +\omega'_{H2} \\ +\omega'_{H3} & 0 & -\omega'_{H1} \\ -\omega'_{H2} & +\omega'_{H1} & 0 \end{bmatrix}$$

where we have:

$\overline{\omega}'_H = (e/m')\overline{B}'_o = (\omega'_{H1}, \omega'_{H2}, \omega'_{H3})$

$\overline{B}'_o$ = static magnetic field as viewed in S'

$-e$ = electron charge

m' = electron mass (stationary)

ω' = circular wave frequency ($e^{+i\omega' t'}$)

ν' = mean collision frequency

N' = electron volume density

$\overline{\overline{I}}$ = identity matrix

The right-hand side of (11-16) can be found explicitly in Ratcliffe [1959].

It follows from (11-13) - (11-14) and the Minkowski expressions (2-52) - (2-53) that:

$$\overline{D} = \epsilon_o \overline{E} \tag{11-17}$$

$$\overline{B} = \mu_o \overline{H}. \tag{11-18}$$

The results (11-17) - (11-18) are the consequence of the Lorentzian viewpoint, that the permittivity and the permeability of the medium are unchanged. The transformations for $\bar{J}'$ and $\bar{E}'$ in (11-15) can be written [Sommerfeld, 1964a] in the form:

$$\bar{J}' = \bar{\bar{\theta}} \cdot (\bar{J} - \rho \bar{v}_o) \tag{11-19}$$

$$\bar{E}' = \bar{\bar{\gamma}} \cdot (\bar{E} + \mu_o \bar{v}_o \times \bar{H}) \tag{11-20}$$

where we define:

$$\bar{\bar{\theta}} = \begin{bmatrix} 1 & 0 & 0 \\ 0 & 1 & 0 \\ 0 & 0 & \gamma \end{bmatrix}, \quad \bar{\bar{\gamma}} = \begin{bmatrix} \gamma & 0 & 0 \\ 0 & \gamma & 0 \\ 0 & 0 & 1 \end{bmatrix}$$

and take:

$$\bar{v} = (0,0,v_o) \quad \gamma = (1 - \beta^2)^{-1/2}, \quad \beta = \beta_L = v_o/c.$$

Substituting (11-19) and (11-20) in (11-15), one has:

$$\bar{\bar{\theta}} \cdot (\bar{J} - \rho \bar{v}) = \bar{\bar{\sigma}}' \cdot \bar{\bar{\gamma}} \cdot (\bar{E} + \mu_o \bar{v}_o \times \bar{H}). \tag{11-21}$$

For clarity, we shall use tensor notations in the following:

$$\theta_{pj}(J_j - \rho v_{oj}) = \sigma'_{pj} \gamma_{jk}(E_k + \mu_o \epsilon_{k\ell m} v_{o\ell} H_m) \tag{11-22}$$

where $\epsilon_{k\ell m}$ is the Levi-Civita symbol [Aris, 1962] and the Einstein summation convention is implied for repeated indices.

Equation (11-22) can be rewritten as follows:

$$J_i = \eta_{ik}(E_k + \mu_o \epsilon_{k\ell m} v_{o\ell} H_m) + \rho v_{oi} \tag{11-23}$$

where we have:

$$\eta_{ik} = \theta^{cong}_{ip} \sigma'_{pm} \gamma_{mk} = \begin{bmatrix} \gamma\sigma'_{11} & \gamma\sigma'_{12} & \sigma'_{13} \\ \gamma\sigma'_{21} & \gamma\sigma'_{22} & \sigma'_{23} \\ \sigma'_{31} & \sigma'_{32} & \frac{1}{\gamma}\sigma'_{33} \end{bmatrix}$$

and

$$\theta^{cong}_{ip} \theta_{pj} = \delta_{ij}.$$

It must be noted that in (11-19) from the continuity relation for $e^{+i\omega t}$:

$$\frac{\partial J_\ell}{\partial x_\ell} = -i\omega\rho. \tag{11-24}$$

The Maxwell curl equations can be written in the following form:

$$\epsilon_{j\ell k}\frac{\partial E_k}{\partial x_\ell} = -i\omega\mu_o H_j - M_j^S \tag{11-25}$$

$$\epsilon_{\ell rj}\frac{\partial H_j}{\partial x_r} = i\omega\epsilon_o E_\ell + J_\ell + J_\ell^S \tag{11-26}$$

where J_ℓ is the convection current in the S system. The divergence of (11-26) gives:

$$\frac{\partial J_\ell}{\partial x_\ell} = -i\omega\rho = -i\omega\epsilon_o \frac{\partial E_\ell}{\partial x_\ell} - \frac{\partial J_\ell^S}{\partial x_\ell} \tag{11-27}$$

thus one has:

$$\rho = \epsilon_o\frac{\partial E_\ell}{\partial x_\ell} + \frac{1}{i\omega} \frac{\partial J_\ell^S}{\partial x_\ell} \tag{11-28}$$

Substituting (11-28) in (11-21) and the result in (11-28), one obtains:

$$\epsilon_{irj}\frac{\partial H_j}{\partial x_r} = i\omega\epsilon_o E_i + \eta_{ik}(E_k + \mu_o\epsilon_{k\ell m}v_{o\ell}H_m) + $$

$$+ \left(\epsilon_o\frac{\partial E_\ell}{\partial x_\ell} + \frac{1}{i\omega} \frac{\partial J_\ell^S}{\partial x_\ell}\right)v_{oi} + J_i^S \tag{11-29}$$

Eliminating H_j from (11-25) and (11-29), one may obtain the electric wave equation in the following form:

$$f_{ij}E_j = a_{ij}J_j^S + b_{ij}M_j^S \tag{11-30}$$

where the tensor differential operators f_{ij}, a_{ij}, and b_{ij} are given by:

$$f_{ij} = \frac{\partial^2}{\partial x_i \partial x_j} - \delta_{ij}\nabla^2 - k_o^2\delta_{ij} + i\omega\mu_o\eta_{ij} + i\omega\mu_o\epsilon_o v_{oi}\frac{\partial}{\partial x_j} -$$

$$- \mu_o\eta_{ik}v_{oj}\frac{\partial}{\partial x_k} + \mu_o\eta_{ij}v_{o\ell}\frac{\partial}{\partial x_\ell}$$

$$a_{ij} = -i\omega\mu_o\delta_{ij} - \mu_o v_{oi}\frac{\partial}{\partial x_j}$$

$$b_{ij} = \mu_o\eta_{ik}\epsilon_{k\ell j}v_{o\ell} - \epsilon_{irj}\frac{\partial}{\partial x_r}\ .$$

Equation (11-30) is the inhomogeneous electric wave equation to be solved for the radiation electric field E_j. Let us define the Fourier integral transform pair as follows:

$$E_j(x_i) = \tau\int_{-\infty}^{\infty} E^j(n_i)e^{-ik_o n_i x_i}dn_i \tag{11-31}$$

$$E^j(n_i) = \tau\int_{-\infty}^{\infty} E_j(x_i)e^{+ik_o n_i x_i}dx_i \tag{11-32}$$

where we take:

$$\tau = \left(\frac{1}{2\pi}\right)^{3/2}$$

an integration in 3 dimensions being understood, $n_i = (n_1, n_2, n_3)$ is the wave refractive index vector, and k_o is the free space wave number.

Defining a similar transformation pair for the currents and substituting these definitions in (11-30), one obtains after performing differentiations under the integration signs:

$$f^{ij}E^j = a^{ij}J_S^j + b^{ij}M_S^j\ , \tag{11-33}$$

where we define:

$$f^{ij} = -k_o^2 n_i n_j + \delta_{ij}k_o^2 n_\ell n_\ell - k_o^2\delta_{ij} +$$

$$+ i\omega\mu_o\eta_{ij} + \omega\mu_o\epsilon_o v_{oi}k_o n_j +$$

$$+ i\mu_o\eta_{ik}v_{oj}k_o n_k - i\eta_{ij}v_{o\ell}k_o n_\ell$$

$$a^{ij} = -i\omega\mu_o\delta_{ij} + i\mu_o k_o v_{oi}n_j$$

$$b^{ij} = \mu_o\eta_{ik}\epsilon_{k\ell j}v_{o\ell} + ik_o\epsilon_{irj}n_r$$

J_S^i and M_S^j are the Fourier transforms of current sources. Equation (11-33) is the algebraic equation for the Fourier transformed components, derived from the wave equation (11-30).

Let F^{ij} be the cofactor matrix of f^{ij}, then one obtains [Aris, 1962]:

$$f^{pj}F^{p\ell} = f\delta_{j\ell} \tag{11-34}$$

where we have:

$$f = \det(f^{p\ell})$$

Then, the solution of (11-33) can be written as

$$E^\ell = \frac{1}{f}[F^{r\ell}a^{rj}J_S^j + F^{r\ell}b^{rj}M_S^j]. \tag{11-35}$$

Defining the scalar Green's function:

$$G(x_i,x_i') = G(x_i',x_i) = \tau^2\int_{-\infty}^{\infty}\frac{1}{f}\,e^{-ik_o n_i(x_i - x_i')}dn_i \tag{11-36}$$

The electric field $E_\ell(x_i)$ can be found from [Unz, 1963]:

$$\begin{aligned} E_\ell(x_i) &= \tau\int_{-\infty}^{\infty}\frac{1}{f}[F^{r\ell}a^{rj}J_S^j + F^{r\ell}b^{rj}M_S^j]\,e^{-ik_o n_i x_i}dn_i \\ &= F_{r\ell}a_{rj}\int_V J_j^S(x_i')G(x_i,x_i')dx_i' + \\ &+ F_{r\ell}b_{rj}\int_V M_j^S(x_i')G(x_i,x_i')dx_i'. \end{aligned} \tag{11-37}$$

Equation (11-37) is the solution for the wave equation (11-30), where $F_{i\ell}a_{ij}$ and $F_{i\ell}b_{ij}$ are tensor differential operators. The primed quantities in (11-36) and (11-37) should not be confused with the primed moving system.

In the tensor η_{ij} one needs the transformation of the plasma parameters from the rest frame. It has been shown in section 2.4 that plasma frequency

$\omega_p = [Ne^2/m\epsilon_o]^{1/2}$ is the Lorentz' invariant, and that $\nu' = \gamma\nu$; the wave frequency will transform, by taking the Doppler effect into account, as $\omega' = \gamma\omega(1 - n_i \beta_i)$ where $\beta_i = (0,0,\beta)$; the gyrofrequency will transform as a result of the transformations for $\bar{B}_o$ and m, in the following form:

$$\omega'_{H1} = \gamma^2 \omega_{H1}; \quad \omega'_{H2} = \gamma^2 \omega_{H2}; \quad \omega'_{H3} = \gamma\omega_{H3} \qquad (11\text{-}38)$$

XII

THE POYNTING VECTOR IN MOVING PLASMAS

12.1 Introduction

The Poynting vector $\overline{S} = \overline{E} \times \overline{H}$ and the Poynting theorem are well understood for stationary media in different formulations [Stratton, 1941; Fano, Chu, and Adler, 1960]. However, the issue has been controversial for the case of moving media. It stems from the Minkowski theory of moving media, which is based on the principle of relativity-the covariance of the physical laws, in our case the Maxwell equations, in systems moving uniformly with respect to each other. The Minkowski theory has been criticized on two major counts. The first one is the asymmetry present in the Minkowski energy-momentum tensor, or stress tensor. The long standing objection to this tensor, on the ground that this asymmetry indicates an unbalanced angualr momentum, was overruled by Beck [1953]. He showed that if matter is properly taken into account along with the field, then one obtains a symmetric tensor. However, Tang and Meixner [1961] considered the whole question to be superficial, from the point of view of relativistic thermodynamics. They obtained a total energy-momentum tensor for the field plus matter, and showed that the oscillating part of it can be reduced to the one obtained by Beck [1953]. The expression for the total tensor, which is symmetric, given by Tang and Meixner [1961], is a sum of the Minkowski tensor and a tensor pertaining to matter. A symmetric tensor was obtained earlier by Abraham [Pauli, 1958], which was based on the Hertz tensor for stationary media, rather than the Maxwell-Heaviside tensor, as was the case for Minkowski. The fact remains that the Minkowski energy-momentum tensor is asymmetric.

The second major criticism of the Minkowski theory is that no reasonable interpretation of the terms $\overline{E} \cdot \partial\overline{D}/\partial t$ and $\overline{H} \cdot \partial\overline{B}/\partial t$ in the Poynting theorem, derived from the Maxwell equations, can be obtained for moving media [Meyers, 1958]. The power converted into the mechanical form, for example, is lost somewhere in the usual Poynting theorem. Considering the fact that a law of physics can only be regarded as invariant in form and meaning for all inertial observers, if it agrees with experiments in a non-proper, moving frame of reference, the Maxwell equations seem to represent a degenerate form of more basic relationships, which are valid in all inertial frames. These more basic relationships were given by the Chu equations [Meyers, 1958; Fano, Chu, and Adler, 1960], which reduce to the Maxwell equations in the proper, rest coordinate system. The Poynting theorem, based on the Chu equations, lead to a reasonable physical interpretation. Furthermore, the energy-momentum tensor, obtained by the Chu equations, is symmetric. The important significance of the Chu equations is that they involve only the two field vectors $\overline{E}$ and $\overline{H}$ and the two material vectors $\overline{P}$ and $\overline{M}$, the polarization and the magnetization densities, respectively. For moving plasmas, Brandstatter [1963] gives a Poynting theorem, which corresponds to the theorem of the Chu theory. Brandstatter uses two field vectors, $\overline{E}$ and $\overline{H}$, and the convection current model for the plasma, obeying the free space Maxwell-Lorentz equation. It will be shown in section 12.3 that the two theorems are identical.

Koyama [Koyama, Chawla, and Unz, 1967] has recently pointed out that when the Minkowski theory is applied to moving plasmas, one obtains two different Poynting vectors, for the polarization current model and for the convection current model. This is not surprising, since the convection current model of the Minkowski theory, as will be shown, is identical with the polarization current model of the Chu theory. But the latter is different from the polarization current model of the Minkowski theory. Hence, the two Poynting vectors, obtained by the application of the Minkowski theory to the two models of the plasma, will be different, since the vector $\overline{H}_c$ of the Chu theory is equal to $\overline{H}_M - \overline{V} \times \overline{P}$ in the Minkowski theory, and $\overline{E}_c$ is equal to $\overline{E}_M$ for non-magnetizable media, as was pointed out by Chawla and Unz [Koyama, Chawla, and Unz, 1967]. A question now arises as to which of the two vectors $\overline{E}_c \times \overline{H}_c$ and $\overline{E}_M \times \overline{H}_M$ is the correct one, and which one needs to be modified. The question can be answered satisfactorily in terms of a meaningful physical interpretation of the Poynting theorem. However, as has been already shown by Meyers [1958], the Poynting theorem, derivable from the Maxwell-Minkowski equations, is not reasonably interpretable for a moving medium. What we then need to do is to modify the Poynting theorem thus obtained, so that it leads to the same physical interpretation as the Poynting theorem derived by the Chu theory. Doing this, as we shall see, amounts to the Poynting theorem, derivable from the amperian formulation, which takes into account the polarization and the magnetization currents. We shall conclude that the Poynting vector of the Minkowski theory should be $\bar{E}_M \times \bar{B}_M/\mu_0$. It will be further shown that one obtains the same Poynting theorem by using small signal theory for drifting plasma. In the next section we shall briefly review the work of Koyama, Chawla, and Unz, [1967]; and in section 12.3 we shall proceed with the corresponding Poynting theorems derivabl from different theories, and obtain the Poynting vectors, so that all the theorems have identical physical interpretation.

12.2 The Minkowski Theory

Let an infinite, isotropic plasma be moving in the $\hat{z}$ direction with the velocity v_0 with respect to the laboratory system, and an electromagnetic wave propagating in the $\hat{x}$ direction. Consider the polarization current model, that is, the plasma is characterized in its rest frame by a dielectric constant:

$$\epsilon = \epsilon_0[1 - \omega_p^2/\gamma^2(\omega - \bar{k}\cdot\bar{v}_0)^2] = \epsilon_0\epsilon_p \tag{12-1}$$

where

$$\bar{k} = k\hat{x},\ \bar{v}_0 = v_0\hat{z}.$$

The Maxwell-Minkowski equations in the laboratory system of coordinates are then:

$$\bar{k}\times\bar{E} = \omega\bar{B} \tag{12-2}$$

$$\bar{k}\times\bar{H} = -\omega\bar{D} \tag{12-3}$$

$$\overline{B} - \frac{1}{c^2}\overline{v}_o \times \overline{E} = \mu_o(\overline{H} - \overline{v}_o \times \overline{D}) \tag{12-4}$$

$$\overline{D} + \frac{1}{c^2}\overline{v}_o \times \overline{H} = \epsilon_o \epsilon_p(\overline{E} + \overline{v}_o \times \overline{B}) \tag{12-5}$$

From (12-2) - (12-5), one obtains:

$$E_x = \frac{\omega v_o(1 - \epsilon_p)}{k(1 - \beta^2)c^2} E_z$$

$$H_x = \frac{(1 - \epsilon_p)v_o}{\mu_o(1 - \beta^2)c^2} E_y; \; H_y = -\frac{\omega\epsilon_o\epsilon_p}{k} E_z; \; H_z = \frac{k}{\omega\mu_o} E_y \tag{12-6}$$

Equation (12-6) leads to:

$$c^2k^2 = \omega^2 - \omega_p^2, \tag{12-7}$$

and taking $\overline{D} = \epsilon_o\overline{E} + \overline{P}$, one obtains:

$$\overline{P} = -\epsilon_o\frac{(1 - \epsilon_p)}{(1 - \epsilon_p\beta^2)} [\mu_o v_o(\hat{y}H_x - \hat{x}H_y) + \overline{E} - \epsilon_p \beta^2 E_z \hat{z}] \tag{12-8}$$

Taking the Poynting vector in the form:

$$\overline{S} = \frac{1}{2} \text{Re}(\overline{E} \times \overline{H}^*) \tag{12-9}$$

by the substitution of (12-6) - (12-7) in (12-9), one obtains:

$$S_{px} = \frac{1}{2} \frac{k}{\omega\mu_o} [E_x^2 + (1 + \frac{v_o^2}{c^2} \frac{\omega_p^2}{\omega^2 - \omega_p^2})E_z^2]; \; S_{py} = 0;$$

$$S_{pz} = -\frac{\epsilon_o v_o}{2} \frac{\omega_p^2}{\omega^2} [E_y^2 + (1 + \frac{v_o^2}{c^2} \frac{\omega_p^2}{\omega^2 - \omega_p^2})E_z^2] \tag{12-10}$$

The Poynting vector $\overline{S}_p$ of the polarization current model, given by (12-10), represents the Poynting vector of an electromagnetic wave, propagating at right angles to the direction of the motion of the plasma. It is noticed that $\overline{S}_p \to \infty$ as $\omega \to \omega_p$.

Consider now the convection current model, that is, the plasma is characterized in its rest system by conductivity

$$\sigma = i\,\epsilon_o \frac{\omega_p^2}{(\omega - \bar{k}\cdot\bar{v}_o)} \tag{12-11}$$

The Maxwell-Lorentz equations in the laboratory system of coordinates may be written as:

$$\bar{k} \times \bar{E} = \omega\,\mu_o \bar{H} \tag{12-12}$$

$$\bar{k} \times \bar{H} = -\omega\epsilon_o \bar{E} - i\bar{J} \tag{12-13}$$

$$\bar{J} = \gamma\sigma[\bar{E} + \mu_o \bar{v}_o \times \bar{H} - \frac{\bar{v}_o}{c}(\frac{\bar{v}_o}{c}\cdot\bar{E})] + i\epsilon_o \bar{v}_o(\bar{k}\cdot\bar{E}) \tag{12-14}$$

Equations (12-11) - (12-14) give the following:

$$E_x = \frac{\omega v_o(1 - \epsilon_p)}{k(1 - \beta^2)c^2} E_z$$

$$H_x = 0;\; H_y = -\frac{k}{\omega\mu_o} E_z;\; H_z = \frac{k}{\omega\mu_o} E_y. \tag{12-15}$$

The substitution of (12-15) in (12-9) gives:

$$S_{cx} = \frac{1}{2}\frac{k}{\omega\mu_o}(E_y^2 + E_z^2);\; S_{cy} = -\frac{1}{2}\mathrm{Re}\;\epsilon_o \frac{\omega_p^2}{\omega^2} v_o E_y^* E_z;$$

$$S_{cz} = -\frac{1}{2}\epsilon_o v_o \frac{\omega_p^2}{\omega^2} E_z^2 \tag{12-16}$$

The Poynting vector $\bar{S}_c$ of the convection current model, given by (12-16), is different from $\bar{S}_p$ of the polarization model, given by (12-10), having the undesirable feature that $S_y \neq 0$. However, the $\bar{H}$ fields in (12-6) and (12-15) are related through the equation [Tai, 1965a]:

$$\bar{H}_c = \bar{H}_p - \bar{v}_o \times \bar{P} \tag{12-17}$$

where the subscript c refers to (12-15), and the subscript p to (12-6). Hence, the $\bar{S}_p$ and $\bar{S}_c$ can be obtained from each other by using (12-17), since $\bar{E}_p = \bar{E}_c$. The relation (12-17) is identical to the relation [Tai, 1964a]:

$$\bar{H}_c = \bar{H}_M - \bar{v}_o \times \bar{P} \tag{12-18}$$

between the $\bar{H}$ vectors of the Chu theory and the Minkowski theory, respectively. This implies that the Poynting vector obtained by the Chu theory should be identical to $\overline{S}_c$. We shall show in the next section that this is indeed the case. The conceptual difference between the two is that in the convection current model, free space is assumed to be pervaded by convective currents, and in the polarization current model of the Chu theory, the free space is assumed to be pervaded by polarization currents. In the next section we shall show that the convection current model of the Minkowski theory, the polarization current model of the Chu theory, and the small signal theory for drifting plasma yield the same set of Maxwell equations, the same Poynting theorems, and hence the same Poynting vector. On the basis that the Poynting theorem thus obtained yields a reasonable physical interpretation, while the Poynting theorem obtained by the Maxwell-Minkowski equation does not [Meyers, 1958] and on the basis of various other arguments against the Minkowski theory [Fano, Chu, and Adler, 1960], we shall discard $\overline{E}_M \times \overline{H}_M$ as the Poynting vector representing the power flow, and modify the equations, to obtain a consistent Poynting theorem and the corresponding Poynting vector.

12.3 Poynting Theorems in Various Formulations

a. The Convection Current Model of the Minkowski Thoery:

For the plasma moving in the $\hat{z}$ direction, the Maxwell equations for this model are:

$$\nabla \times \overline{E} = -\mu_0 \frac{\partial \overline{H}}{\partial t} \tag{12-19}$$

$$\nabla \times \overline{H} = \epsilon_0 \frac{\partial \overline{E}}{\partial t} + \overline{J}_c \tag{12-20}$$

where one has:

$$\overline{J}_c = \overline{\overline{\eta}} \cdot (\overline{E} + \mu_0 \overline{v}_0 \times \overline{H}) + \rho \overline{v}_0 \tag{12-21}$$

$$\overline{\overline{\eta}} = \begin{bmatrix} \gamma & 0 & 0 \\ 0 & \gamma & 0 \\ 0 & 0 & \frac{1}{\gamma} \end{bmatrix} \sigma, \quad \rho = \epsilon_0 (\nabla \cdot \overline{E}).$$

Taking the dot product of (12-19) with $\overline{H}$, the dot product of (12-20) with $\overline{E}$, and subtracting the latter from the former, one obtains:

$$-\nabla \cdot (\overline{E} \times \overline{H}) - \frac{\partial}{\partial t} \frac{1}{2} (\epsilon_o |\overline{E}|^2 + \mu_o |\overline{H}|^2) = +\overline{E} \cdot \overline{J}_c \qquad (12\text{-}22)$$

Upon integrating over a fixed volume V, the Poynting theorem (12-22) leads to a physical interpretation which is well known [Brandstatter, 1963]. The Poynting vector $\overline{E} \times \overline{H}$ represents the electromagnetic power density flowing out of the volume through its bounding surface; the negative sign in front of it represents the power density flowing into the volume.

b. The Polarization Current Model of the Chu Theory:

The Chu equations for this model are:

$$\nabla \times \overline{E}_p = -\mu_o \frac{\partial \overline{H}_p}{\partial t} \qquad (12\text{-}23)$$

$$\nabla \times \overline{H}_p = \epsilon_o \frac{\partial \overline{E}_p}{\partial t} + \overline{J}_p \qquad (12\text{-}24)$$

where $\overline{J}_p = \frac{\partial \overline{P}}{\partial t} + \nabla \times (\overline{P} \times \overline{v}_o)$

The Poynting theorem derived from (12-23) - (12-24) is

$$-\nabla \cdot (\overline{E}_p \times \overline{H}_p) - \frac{\partial}{\partial t} \frac{1}{2}(\epsilon_o |\overline{E}_p|^2 + \mu_o |\overline{H}_p|^2) = \overline{E}_p \cdot \overline{J}_p \qquad (12\text{-}25)$$

Equations (12-22) and (12-25) are identical, if we can show that $J_c = J_p$. The right-hand sides of (12-22) and (12-25) represent the total electromagnetic power, transformed within V to some other form of energy, through the motion of the electric charges, in the presence of an electromagnetic field. Part of the electromagnetic power converted is identifiable as work per unit time, done by the macroscopic electromagnetic forces acting on each grain of matter; the rest of it represents the energy stored or dissipated in the matter [Fano, Chu, and Adler, 1960].

Consider the current density in the rest system of the plasma (the primed system):

$$\overline{J}' = \sigma \overline{E}'_p = \frac{\partial \overline{P}'}{\partial t'} = \gamma(\frac{\partial}{\partial t} + \overline{v}_o \cdot \nabla)\overline{P}' \qquad (12\text{-}26)$$

Writing $\overline{P}' = (\epsilon - \epsilon_0)\overline{E}'_p$, (12-26) yields the operator:

$$\sigma = \gamma(\epsilon - \epsilon_0)\left(\frac{\partial}{\partial t} + \overline{v}_0 \cdot \nabla\right). \tag{12-27}$$

Now $\overline{J}_p$ in (12-24) may be rewritten as:

$$\overline{J}_p = \left(\frac{\partial}{\partial t} + \overline{v}_0 \cdot \nabla\right)\overline{P} - \overline{v}_0(\nabla \cdot \overline{P}) \tag{12-28}$$

Using the constitutive relation [Tai, 1967]:

$$\overline{P} = (\epsilon - \epsilon_0)\overline{\overline{\gamma}} \cdot \overline{\overline{\gamma}} \cdot (\overline{E}_p + \mu_0 \overline{v}_0 \times \overline{H}_p) \tag{12-29}$$

where we define:

$$\overline{\overline{\gamma}} = \begin{bmatrix} \gamma & 0 & 0 \\ 0 & \gamma & 0 \\ 0 & 0 & 1 \end{bmatrix}, \quad \gamma = (1 - \beta^2)^{-1/2}$$

and having the relation:

$$\nabla \cdot \overline{P} = -\epsilon_0 \nabla \cdot \overline{E}_p = -\rho$$

obtained by taking divergence of (12-24), equation (12-28) becomes, by using (12-21):

$$\begin{aligned} \overline{J}_p &= (\epsilon - \epsilon_0)\left(\frac{\partial}{\partial t} + \overline{v}_0 \cdot \nabla\right)\overline{\overline{\gamma}} \cdot \overline{\overline{\gamma}} \cdot (\overline{E}_p + \mu_0 \overline{v}_0 \times \overline{H}_p) + \rho \overline{v}_0 \\ &= \overline{\overline{\eta}} \cdot (\overline{E}_p + \mu_0 \overline{v}_0 \times \overline{H}_p) + \rho \overline{v}_0 = \overline{J}_c \end{aligned} \tag{12-30}$$

Therefore, the Poynting vectors obtained for the above two models are identical to each other.

c. The Small Signal Theory:

The Maxwell equations of the small signal theory are [Oster, 1960]:

$$\nabla \times \overline{E} = -\mu_0 \frac{\partial \overline{H}}{\partial t} \tag{12-31}$$

$$\nabla \times \overline{H} = \epsilon_0 \frac{\partial \overline{E}}{\partial t} + \overline{J}_s \tag{12-32}$$

where we have:

$$\overline{J}_s = -e(N_o\overline{v}_s + N\overline{v}_o). \tag{12-33}$$

In the above, N and $\overline{v}_s$ are the perturbed electron density and the perturbed velocity, respectively, and the zero subscript quantities are their stationary average values.

Equations (12-31) - (12-32) give the Poynting theorem in the form:

$$-\nabla \cdot (\overline{E} \times \overline{H}) - \frac{\partial}{\partial t}\frac{1}{2}(\epsilon_o|\overline{E}|^2 + \mu_o|\overline{H}|^2) = \overline{E} \cdot \overline{J}_s \tag{12-34}$$

which are identical to (12-22) and to (12-25), if we can show that $\overline{J}_s = \overline{J}_p = \overline{J}_c$.

In the rest system of the plasma, one has:

$$\overline{J}' = \frac{\partial \overline{P}'}{\partial t'} = \gamma(\frac{\partial}{\partial t} + \overline{v}_o \cdot \nabla)\overline{P}' = -eN_o'\overline{v}' \tag{12-35}$$

By using in (12-35) the transformations $\overline{P} = \overline{\overline{\gamma}} \cdot \overline{P}'$ [Tai, 1967], $N_o' = N_o/\gamma$, and $\overline{v}' = \gamma(\overline{\overline{\alpha}} \cdot \overline{v} - \gamma\overline{v}_o)$ [Moller, 1952], where:

$$\overline{\overline{\alpha}} = \begin{bmatrix} 1 & 0 & 0 \\ 0 & 1 & 0 \\ 0 & 0 & \gamma \end{bmatrix}$$

and the definition $\overline{v} = \overline{v}_s + \overline{v}_o$ so that $\overline{v}' = \overline{\overline{\alpha}} \cdot \overline{v}_s$, one obtains:

$$-eN_o\overline{v}_s = (\partial/\partial t + \overline{v}_o \cdot \nabla)\overline{P}. \tag{12-36}$$

The continuity equation is given in the form:

$$\nabla \cdot (N\overline{v}_o + N_o\overline{v}_s) = -\frac{\partial N}{\partial t} \tag{12-37}$$

From (12-36) and (12-37) one obtains:

$$-eN\overline{v}_o = -(\nabla \cdot \overline{P})\overline{v}_o \tag{12-38}$$

From (12-36) and (12-38) one has:

$$\overline{J}_s = -eN_o\overline{v}_s - eN\overline{v}_o = (\frac{\partial}{\partial t} + \overline{v}_o \cdot \nabla)\overline{P} - \overline{v}_o(\nabla \cdot \overline{P}) =$$

$$= \frac{\partial}{\partial t} + \nabla \times (\overline{P} \times \overline{v}_o) = \overline{J}_p \tag{12-39}$$

Hence, the Poynting theorems arrived at from the above three theories are identical, and so are the corresponding Poynting vectors.

d. The Polarization Current Model of the Minkowski Theory:

The Maxwell equations for the moving plasma in this model are:

$$\nabla \times \overline{E}_M = - \frac{\partial \overline{B}}{\partial t} \tag{12-40}$$

$$\nabla \times \overline{H}_M = \frac{\partial \overline{D}}{\partial t} \tag{12-41}$$

and the Poynting theorem derived from these equations is given by:

$$\nabla \cdot (\overline{E}_M \times \overline{H}_M) + (\overline{H}_M \cdot \frac{\partial \overline{B}}{\partial t} + \overline{E}_M \circ \frac{\partial \overline{D}}{\partial t}) = 0 \tag{12-42}$$

where $\overline{B}$ and $\overline{D}$ are given in terms of $\overline{E}$ and $\overline{H}$ by (12-4) - (12-5). Clearly, (12-42) is different from (12-22), and no amount of juggling of terms, or Lorentz transforming back and forth between frames, leads to a reasonable interpretation of the terms $\overline{E}_M \cdot \frac{\partial \overline{D}}{\partial t}$ and $\overline{H}_M \cdot \frac{\partial \overline{B}}{\partial t}$. On further argument that the Minkowski theory does not predict electric forces on nonconducting uncharged dielectric bodies, it is evident that (12-42) needs to be modified, so as to obtain the Poynting theorem (12-25).

Writing $\overline{D} = \epsilon_o \overline{E} + \overline{P}$ in (12-41) and adding $-\nabla \times (\overline{v}_o \times \overline{P})$ on both sides of it, one has:

$$\nabla \times (\overline{H}_M - \overline{v}_o \times \overline{P}) = \epsilon_o \frac{\partial \overline{E}}{\partial t} + \frac{\partial \overline{P}}{\partial t} + \nabla \times (\overline{P} \times \overline{v}_o) \tag{12-43}$$

But $\overline{B} = \mu_o(\overline{H}_M + \overline{M})$ and $\overline{M} = -\overline{v}_o \times \overline{P}$; hence, one has:

$$\nabla \times (\overline{B}/\mu_o) = \epsilon_o \frac{\partial \overline{E}}{\partial t} + \frac{\partial \overline{P}}{\partial t} + \nabla \times (\overline{P} \times \overline{v}_o) \tag{12-44}$$

The Poynting theorem derivable from (12-40) and (12-44) is then:

$$-\nabla \cdot (\overline{E}_M \times \overline{B}/\mu_o) - \frac{\partial}{\partial t} \frac{1}{2}(\epsilon_o |\overline{E}|^2 + \mu_o \left|\frac{\overline{B}}{\mu_o}\right|^2) = \overline{E} \cdot (\frac{\partial \overline{P}}{\partial t} + \nabla \times \overline{P} \times \overline{v}_o) \tag{12-45}$$

Equation (12-45) is identical with (12-25) since $\overline{B}/\mu_o = \overline{H}_c$ [Koyama, Chawla, and Unz, 1967]. Hence, the Poynting vector in the Minkowski theory for polarizable moving media should be $\overline{E} \times \overline{B}/\mu_o$, and not $\overline{E} \times \overline{H}$.

It should be noted that (12-45) is a particular case of the Poynting theorem of the Amperian formulation eq. A1.200 [Fano, Chu, and Adler, 1960] for polarizable media only. That theorem was criticized on the basis of the power absorption in magnetized materials; however, for only polarizable media ($\overline{M}' = 0$), such as a plasma, (12-45) agrees completely with the Poynting theorem (12-25) of the Chu formulation.

XIII

PROPAGATION IN TWO STREAM MAGNETO-PLASMA

13.1 Introduction

So far, we have considered wave propagation in a single stream magneto-plasma. However, physical situations occur where there may be more than one stream. In the present chapter we shall consider electromagnetic wave propagation in a two stream magneto-plasma. In the laboratory system of reference, one observes two streams of uniform plasmas of different densities, flowing in two different directions, with uniform velocities, through a static magnetic field. Each of the streams contributes an a.c. electric current density term in the Maxwell equations. This current density is obtained by applying the Lorentz transformations to the current density of the stream, in its rest system, arising from the passage of the electromagnetic wave disturbance [Tai, 1965a; chawla and Unz, 1966a]. The plasma streams in their rest systems are represented by conductivity tensors, thus the model considered there is that of free space pervaded by the convective currents, arising from the electromagnetic wave disturbance of a uniformly moving plasma stream. Our object here is to obtain the dispersion relation for the electromagnetic wave [Chawla and Unz, 1969c].

The dispersion relation for this general situation is given in a determinantal equation form. As will be seen, this determinant is hideous to expand, in order to obtain an algebraic equation for this general case. The complications arising for an n-stream magneto-plasma can very well be imagined. The determinant is expanded for the particular case of one of the streams being stationary in the laboratory system, or in the rest system of one of the streams, and the direction of the flow, the wave propagation, and the static magnetic field are the same, namely, the $\hat{z}$ direction. It is shown that for the particular case of only one stream, the dispersion equation reduces to the previously found result. The dispersion equations found here may be used to study amplification and instabilities [Getmantsev and Rapoport, 1960; Briggs, 1964]; this, however, does not fall under the subject of the present book. In the subsequent sections tensor notation is used [Aris, 1962].

13.2 The General Wave Equation

Let two infinite plasma streams, with uniform, unperturbed electron densities N_o^u and N_o^v, and described by their conductivity tensors [Brandstatter, 1963] σ_{ij}^u and σ_{ij}^v in their rest system be moving uniformly with velocities u_i and v_i, respectively, in relation to a laboratory observer, through a static magnetic field B_{oi}. The electromagnetic disturbance in such a system must satisfy the following Maxwell equations, where harmonic time variation $e^{i\omega t}$ is assumed and suppressed:

$$\epsilon_{jk\ell}E_{\ell,k} = -i\omega\mu_o H_j \tag{13-1}$$

$$\epsilon_{irj}H_{j,r} = i\,\omega\,\epsilon_o E_i + J_i^u + J_i^v \tag{13-2}$$

where J_i^u and J_i^v are the electric current density contributions of the two streams, given by [Chawla and Unz, 1966a]:

$$J_i^u = \eta_{ik}^u E_k + \zeta_{ik}^u \gamma^u \mu_o \epsilon_{k\ell m} u_\ell H_m + \gamma^u \rho^u \theta_{ik}^u u_k \tag{13-3}$$

$$J_i^v = \eta_{ik}^v E_k + \zeta_{ik}^v \gamma^v \mu_o \epsilon_{k\ell m} v_\ell H_m + \gamma^v \rho^v \theta_{ik}^v v_k \tag{13-4}$$

where ϵ_{ijk} is the Levi-Civita symbol [Aris, 1962], and

$$\eta_{ik}^u = \theta_{ij}^u \sigma_{j\ell}^u \gamma_{\ell k}^u$$

$$\theta_{ij}^u = \left[\delta_{ij} + (\gamma^u - 1)\frac{u_i u_j}{u^2}\right]^{-1}$$

$$\gamma_{\ell k}^u = \gamma^u \delta_{\ell k} + (1 - \gamma^u)\frac{u_\ell u_k}{u^2}$$

$$\zeta_{ik}^u = \theta_{ij}^u \sigma_{jk}^u$$

$$\gamma^u = (1 - u^2/c^2)^{1/2};\ u^2 = u_1^2 + u_2^2 + u_3^2;$$

the corresponding v superscript quantities follow from the above in the same form. From (13-3) and (13-4) one obtains:

$$J_i^u + J_i^v = (\eta_{ik}^u + \eta_{ik}^v)E_k + \mu_o \epsilon_{k\ell m}(\gamma^u \zeta_{ik}^u u_\ell + \gamma^v \zeta_{ik}^v v_\ell)H_m + \\ + (\gamma^u \rho^u \theta_{ik}^u u_k + \gamma^v \rho^v \theta_{ik}^v v_k) \tag{13-5}$$

Using (13-3) in the continuity equation:

$$J_{i,i}^u = -i\,\omega\,\rho^u \tag{13-6}$$

one has:

$$\rho^u = \frac{1}{(-i\omega - \gamma^u \theta_{ik}^u u_k \frac{\partial}{\partial x_i})}[\eta_{ik}^u E_{k,i} + \gamma^u \mu_o \epsilon_{k\ell m}\zeta_{ik}^u u_\ell H_{m,i}] \tag{13-7}$$

Using the continuity equations in the results of taking the divergence of (13-2), one obtains:

$$\rho^v = \epsilon_o E_{k,k} - \rho^u \tag{13-8}$$

Therefore one has:

$$J_i^u + J_i^v = (\eta_{ik}^u + \eta_{ik}^v)E_k + \mu_o \epsilon_{k\ell m}(\gamma^u \zeta_{ik}^u u_\ell + \gamma^v \zeta_{ik}^v v_\ell)H_m + \\ + \gamma^v \epsilon_o E_{k,k}\theta_{i\ell}^v v_\ell - \\ - \frac{1}{(i\omega + \gamma^u \theta_{\ell k}^u u_k \frac{\partial}{\partial x_\ell})}[\eta_{\ell k}^u E_{k,\ell} + \gamma^u \mu_o \epsilon_{k\ell m}\zeta_{pk}^u u_\ell H_{m,p}]\cdot(\gamma^u \theta_{ik}^u u_k - \gamma^v \theta_{ik}^v v_k) \tag{13-9}$$

Substituting (13-9) in (13-2) and eliminating H_j from resulting equation and (13-1), one has:

$$\left\{\frac{\partial^2}{\partial x_i \partial x_k} - \frac{\partial^2}{\partial x_j \partial x_j}\delta_{ik} - k_o^2 \delta_{ik} + \right.$$

$$+ i\,\omega\,\mu_o[\eta_{ik}^u + \eta_{ik}^v - \frac{(\gamma^u \theta_{ij}^u u_j - \gamma^v \theta_{ij}^v v_j)}{(i\omega + \gamma^u \theta_{jk}^u u_k \frac{\partial}{\partial x_j})}\eta_{pk}^u \frac{\partial}{\partial x_p}] -$$

$$-\mu_o[\gamma^u \zeta^u_{ir} u_k + \gamma^v \zeta^v_{ir} v_k - \frac{\gamma^u \theta^u_{ij} u_j - \gamma^v \theta^v_{ij} v_j}{(i\omega + \gamma^u \theta^u_{jk} u_k \frac{\partial}{\partial x_j})} \gamma^u \zeta^u_{pr} u_k] \frac{\partial}{\partial x_r} +$$

$$+\mu_o[\gamma^u \zeta^u_{ik} u_\ell + \gamma^v \zeta^v_{ik} v_\ell - \frac{\gamma^u \theta^u_{ij} u_j - \gamma^v \theta^v_{ij} v_j}{(i\omega + \gamma^u \theta^u_{jk} u_k \frac{\partial}{\partial x_j})} \gamma^u \zeta^u_{pk} u_\ell] \frac{\partial}{\partial x_\ell} +$$

$$\left. + i\gamma^v \omega \epsilon_o \mu_o \theta^v_{i\ell} v_\ell \frac{\partial}{\partial x_k} \right\} E_k = 0 \qquad (13\text{-}10)$$

where $k_o^2 = \omega^2 \mu_o \epsilon_o$. Equation (13-10) is the general wave equation for a wave propagating in an arbitrary direction in a two stream magneto-plasma.

13.3 Particular Cases

When both of the streams flow in the same direction for example the $\hat{z}$ direction, the equation (13-10) reduces to:

$$\left\{ \frac{\partial^2}{\partial x_i \partial x_k} - \frac{\partial}{\partial x_\ell \partial x_\ell} \delta_{ik} - k_o^2 \delta_{ik} + \right.$$

$$+ i\omega\mu_o[\eta^u_{ik} + \eta^v_{ik} - \frac{(u_i - v_i)}{(i\omega + u_j \frac{\partial}{\partial x_j})} \eta^u_{pk} \frac{\partial}{\partial x_p}] -$$

$$- \mu_o[\eta^u_{ir} u_k + \eta^v_{ir} v_k - \frac{(u_i - v_i)}{(i\omega + u_j \frac{\partial}{\partial x_j})} \eta^u_{pr} \frac{\partial}{\partial x_p} u_k] \frac{\partial}{\partial x_r} +$$

$$+ \mu_o[\eta^u_{ik} u_\ell + \eta^v_{ik} v_\ell - \frac{(u_i - v_i)}{(i\omega + u_j \frac{\partial}{\partial x_j})} \eta^u_{pk} \frac{\partial}{\partial x_p} u_\ell] \frac{\partial}{\partial x_\ell} +$$

$$\left. + i\omega\epsilon_o\mu_o v_i \frac{\partial}{\partial x_k} \right\} E_k = 0 \qquad (13\text{-}11)$$

where

$$\eta_{ik}^{u} = \begin{bmatrix} \gamma^{u}\sigma_{11}^{u} & \gamma^{u}\sigma_{12}^{u} & \sigma_{13}^{u} \\ \gamma^{u}\sigma_{21}^{u} & \gamma^{u}\sigma_{22}^{u} & \sigma_{23}^{u} \\ \sigma_{31}^{u} & \sigma_{32}^{u} & \frac{1}{\gamma^{u}}\sigma_{33}^{u} \end{bmatrix}$$

The dispersion relation for a wave with the space factor $\exp(-ik_x x - ik_y y - ik_z z)$ is obtained by substituting this factor in (13-10) or (13-11) and equating the determinant of the coefficient matrix of E_k to zero. As can be seen, this involves a tremendous amount of algebra, which we shall not carry here, but instead we will consider a more particular case.

If one of the streams, say the one with velocity u_i, be stationary ($u_i = 0$), then (13-11) reduces to:

$$\left\{ \frac{\partial^2}{\partial x_i \partial x_k} - \frac{\partial^2}{\partial x_j \partial x_j} \delta_{ik} - k_o^2 \delta_{ik} + i\omega\mu_o \sigma_{ik}^{u} + i\omega\mu_o \eta_{ik}^{v} + \right.$$

$$\left. + \mu_o v_i \sigma_{pk} \frac{\partial}{\partial x_p} - \mu_o \eta_{ir}^{v} v_k \frac{\partial}{\partial x_r} + \mu_o \eta_{ik}^{v} v_j \frac{\partial}{\partial x_j} \right\} E_k = 0 \quad (13\text{-}12)$$

Consider now that the fields vary as $i(\omega t - kz\cos\theta - kx\sin\theta)$; then the elements of the matrix $\bar{\bar{A}}$ in (13-12), $A_{ik}E_k = 0$, are as follows:

$$\bar{\bar{A}}_{11} = -k^2\sin^2\theta + k^2 - k_o^2 + i\omega\mu_o(\sigma_{11}^{u} + \eta_{11}^{v}) - i\mu_o v_o \eta_{11}^{v} k \cos\theta$$

$$A_{12} = i\omega\mu_o(\sigma_{12}^{u} + \eta_{12}^{v}) - i\mu_o v_o \eta_{12}^{v} k \cos\theta$$

$$A_{13} = -k^2 \cos\theta \sin\theta + ik\mu_o v_o \eta_{11}^{v} \sin\theta$$

$$A_{21} = -i\omega\mu_o(\sigma_{12}^{u} + \eta_{12}^{v}) + i\mu_o v_o \eta_{12}^{v} k \cos\theta$$

$$A_{22} = k^2 - k_o^2 + i\omega\mu_o(\sigma_{11}^{u} + \eta_{11}^{v}) - i\mu_o v_o \eta_{11}^{v} k \cos\theta$$

$$A_{23} = -i\mu_o v_o \eta_{12}^{v} k \sin\theta$$

$$A_{31} = -k^2 \cos\theta \sin\theta + \omega\mu_o \epsilon_o v_o k \sin\theta - i\mu_o v_o \sigma_{11}^{u} k \sin\theta$$

$$A_{32} = -i\mu_o v_o \sigma_{12}^{u} k \sin\theta$$

$$A_{33} = -k^2 \cos^2\theta + k^2 - k_o^2 + i\omega\mu_o(\sigma_{33}^{u} + \eta_{33}^{v}) + \omega\mu_o \epsilon_o v_o k \cos\theta -$$

$$-i\mu_o v_o \sigma_{33}^{u} k \cos\theta.$$

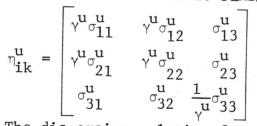

where we take:

$$v_i = (0,0,v_o), B_{oi} = (0,0,B_o),\quad \eta^v_{11} = \eta^v_{22},\quad \eta^v_{21} = -\eta^v_{12},$$
$$\sigma^u_{11} = \sigma^u_{22},\quad \sigma^u_{21} = -\sigma^u_{12}.$$

For the particular case of wave propagation in the $\hat{z}$ direction ($\theta = 0$), A_{ij} reduce to:

$$A_{11} = A_{22} = k^2 - k_o^2 + i\omega\mu_o(\sigma^u_{11} + \eta^v_{11}) - i\mu_o v_o \eta^v_{11} k$$
$$A_{12} = -A_{21} = i\,\omega\mu_o(\sigma^u_{12} + \eta^v_{12}) - i\mu_o v_o \eta^v_{12} k$$
$$A_{13} = A_{23} = A_{31} = A_{32} = 0$$
$$A_{33} = -k_o^2 + i\omega\mu_o(\sigma^u_{33} + \eta^v_{33}) + \omega\mu_o\epsilon_o v_o k - i\mu_o v_o \sigma^u_{33} k$$

The solution of the determinantal equation $|A_{ij}| = 0$ for this case gives $A_{11} = \pm iA_{12}$ or $A_{33} = 0$.

The first solution gives:

$$k^2 - k_o^2 + i\omega\mu_o(\sigma^u_{11} + \eta^v_{11}) - i\,\mu_o v_o \eta^v_{11} k =$$
$$= \pm\,(\omega\sigma^u_{12} + \omega\eta^v_{12} - v_o\eta^v_{12}k)\mu_o \qquad (13\text{-}13)$$

where, neglecting collisions, one has:

$$\sigma^u_{11} = \frac{-iX^u\omega\epsilon_o}{1-(Y^u)^2};\quad \sigma^u_{12} = \frac{-X^uY^u\omega\epsilon_o}{1-(Y^u)^2};\quad \eta^v_{11} = \frac{-iX^v\omega^v\epsilon_o}{1-(Y^v)^2};$$

$$\eta^v_{12} = \frac{-X^vY^v\omega^v\epsilon_o}{1-(Y^v)^2};\qquad \gamma = (1 - v_o^2/c^2)^{-1/2}$$

$$\omega^v = \gamma(\omega - k_o v_o);$$

$$X^u = \frac{e^2N_o^u}{m\epsilon_o\omega^2} = \left(\frac{\omega_p^u}{\omega}\right)^2;\quad X^v = \frac{e^2N_o^v}{m\epsilon_o\gamma^2(\omega - kv_o)^2} = \left(\frac{\omega_p^v}{\gamma(\omega - kv_o)}\right)^2;$$

$$Y^u = \frac{eB_o}{m\omega} = \frac{\omega_B}{\omega};\quad Y^v = \frac{eB_o}{m(\omega - kv_o)} = \frac{\omega_B}{(\omega - kv_o)}$$

Equation (13-13) can be rewritten in terms of the plasma parameters as follows:

$$(\omega^2 \pm \omega\omega_B)(\omega - kv_o \pm \omega_B)(k^2 - k_o^2) + k_o^2(\omega_p^u)^2(\omega - kv_o \pm \omega_B) + $$
$$+ (\omega_p^v)^2 (\omega - kv_o)(\omega^2 \pm \omega\omega_B)/c^2 = 0 \qquad (13\text{-}14)$$

which, for the particular case of only one stream $\omega_p^u = 0$, reduces to

$$(\omega - kv_o \pm \omega_B)(k^2 - k_o^2) + (\omega_p^v)^2(\omega - kv_o)/c^2 = 0. \qquad (13\text{-}15)$$

Equation (13-15) is identical with (3-38) and (6-10) for $Z = 0$ and $n = ck/\omega$:

$$(1 - n\beta \pm Y)(n^2 - 1) + X(1 - n\beta) = 0 \qquad (13\text{-}16)$$

Equation (13-16) agrees with the well-known cubic equation of the drifting magneto-plasma [Unz, 1965a]. It should be noted that (13-14) is also a cubic equation in k and ω. The analysis of (13-14) and (13-15) for instabilities and amplification, could be accomplished, by using the criterion given by Briggs [1964].

XIV

SUMMARY

Various aspects of electromagnetic wave propagation in moving magneto-plasmas have been considered in the present book. They are: The characteristic equations of waves propagating in an arbitrary direction in temperate and warm plasmas, moving uniformly at an arbitrary angle to a uniform static magnetic field; the normal and oblique incidence of linearly polarized, plane electromagnetic waves, on a semi-infinite or a slab of temperate or warm magneto-plasma, moving through free space or a dielectric medium; the radiation from electromagnetic sources in the presence of a uniformly moving temperate magneto-plasma; wave propagation in a two stream magneto-plasma, and the Poynting vector.

The characteristic refractive index equations for the waves in a moving, temperate or warm, magneto-plasma have been found by using two different models, which gave identical results. The characteristic refractive index equations have been found by using two approaches: In the first approach, which applies for non-relativistic velocities only, the equation was obtained in a determinantal form; in the second approach, which applies for relativistic velocities as well, the equation was obtained in an algebraic form, by using Lorentz transformations. It has been noted that for the non-drifting boundary value problem, the characteristic equation for the warm magneto-plasma is of the eighth order, and for the drifting boundary value problem, the characteristic equation is of the sixth order.

The reflections of normally incident, linearly polarized, plane electromagnetic waves, from a semi-infinite temperate magneto-plasma, moving through free space or a dielectric media, have been investigated. It was found that the amplitude of the reflected wave tends to infinity in the limit, when the velocity of the plasma, moving towards the source of the incident wave, tends to the phase velocity of the wave in the stationary free space or dielectric medium, and when the incident wave frequency is equal to half of the cyclotron frequency. The polarization of the reflected wave has been found to be different from the polarization of the incident wave, when the static magnetic field is along the direction of the incident wave. For the magnetic field perpendicular to the direction of the incident wave, the polarization of the reflected wave is the same as the polarization of the incident wave. For the normal incidence of electromagnetic waves on a moving, temperate, magneto-plasma slab, it has been found that with increase in the magneto-static field, the absolute maximum of the reflection coefficient increases for the parameters chosen. An increase in the magneto-static field makes the slab more transparent, particularly in the region where the transmission coefficient for no magneto-static field is very small. A dielectric-like behavior is observed for large magneto-static fields. The sum of the power reflection and the power transmission coefficients is no longer equal to unity for plasma velocities different from zero. The reflection of an electromagnetic wave, obliquely incident on a semi-infinite, temperate, magneto-plasma, moving through free space, has been considered for the parallel and the perpendicular polarization of the incident electromagnetic wave; the formal solution of the general problem of oblique incidence of an electromagnetic wave on a warm, general magneto-plasma slab, moving through a dielectric, has been given.

The formal solution of the problem of the radiation from electromagnetic sources, in the presence of a moving, temperate magneto-plasma, has been given, using the Lorentzian viewpoint, which considers the plasma to be a free space, pervaded by currents, arising from the electromagnetic disturbance. It has been found that the formal solution thus obtained is more straight-forward, compared to the one based on the Maxwellian viewpoint and the Minkowski theory for anisotropic media.

Considerations of the Poynting vector in moving plasmas or in general moving polarizable media, using different formulations, has shown that the interpretation of the Poynting vector as a power flow density should depend on the correct physical interpretation of the Poynting theorem. Based on this and other arguments, the Poynting theorem of the Minkowski theory, as applied to polarizable media, was modified; it has been shown that the correct Poynting vector of this theory, as applied to polarizable media, should be $\bar{E} \times \bar{B}/\mu_0$, and not $\bar{E} \times \bar{H}$. Finally, the dispersion relation for waves propagating in a two stream magneto-plasma has been found and particular cases have been discussed. An extensive Bibliography on the subjects related to the present book is given at the end of the book.

There are numerous physical problems in which it is necessary to take into account the interaction between electromagnetic waves and a moving plasma. Among such problems are the analysis of traveling wave tubes and other microwave electron tubes for the amplification and oscillations of microwaves; the diagnosis of shock waves and moving plasma diagnostics; frequency multiplication by drifting plasma; reflection and transmission of electromagnetic waves through a drifting ionosphere; propagation of whistlers and amplification of very low frequency emissions; Cerenkov radiation; stabilization of the oscillations of a plasma wall; reflection and transmission of electromagnetic waves from a space vehicle upon re-entry; the effect of drifting inhomogeneities in ionospheric propagation; ionospheric backscatter; magneto-spheric boundary effects; the solar corona phenomena; random and irregular ionospheric return; ionospheric propagation effects due to a nuclear explosion including Faraday rotation effects; study of the structure of a traveling ionospheric disturbance including effects on the ray path; effects on cosmic and planetary radio sources'radiation through the moving ionosphere; effects on satellite antennas in the ionosphere with relative motion; the effect of radar scattering cross-section due to relative motion; radio frequency measurements in the ionosphere; propagation losses; electron temperatures in the ionosphere and radio aurora; diffraction and surface waves due to moving bounded ionosphere inhomogeneities; radiation resistance in a bounded streaming plasma in a shock tube; effects on ducting in the troposphere; effects of cosmic rays and solar flares in the ionosphere; effects of acoustic waves in the ionosphere; interference effects in the moving ionosphere; electromagnetic reflections from solar flares; aurorael geomagnetic disturbances in the ionosphere; and numerous others.

Most of the solutions presently available concerning the interaction between electromagnetic waves and moving magneto-plasmas are for the infinite plasma with no boundaries or for finite drifting plasma with non-moving boundaries. A more complete solution of the type of problems described above should include the physical reality in numerous physical phenomena of an electromagnetic wave incident on a moving, bounded magneto-plasma with moving boundaries. The theoretical and numerical solutions of problems involving the interaction between electromagnetic waves and bounded drifting magneto-plasmas with moving boundaries might explain additional physical phenomena and give new interpretations to previous results.

It is hoped that the present book will prove to be useful as a starting point for the analysis of some of the above-mentioned physical phenomena, as well as for other related phenomena.

APPENDIXES

APPENDIX A

Potential Equations in an Isotropic Moving Medium

The potential equations for the relativistic case were derived by Papas [1965], by applying the Lorentz transformation to the equations in the rest frame of the moving medium, and by defining a four-potential for the four-dimensional Maxwell equations. In view of the recent study [Tai, 1964a] of the Minkowski theory, it is of interest to derive the potential equations for the non-relativistic case, based on this theory [Chawla and Unz, 1966c].

The Maxwell equations for a dielectric medium in the laboratory frame of reference are:

$$\nabla \times \bar{E} = - \frac{\partial \bar{B}}{\partial t} \tag{A-1}$$

$$\nabla \times \bar{H} = \bar{J} + \frac{\partial \bar{D}}{\partial t} \tag{A-2}$$

$$\nabla \cdot \bar{D} = \rho \tag{A-3}$$

$$\nabla \cdot \bar{B} = 0 \tag{A-4}$$

The constitutive relations in the non-relativistic case of a moving, homogeneous, isotropic medium, where $\beta^2 = v_o^2/c^2 \ll 1$, may be written as [Tai, 1964 a,b]:

$$\bar{B} = \mu\bar{H} - \bar{\Lambda} \times \bar{E} \tag{A-5}$$

$$\bar{D} = \epsilon\bar{E} + \bar{\Lambda} \times \bar{H} \tag{A-6}$$

where we define:

$$\bar{\Lambda} = (\epsilon\mu - \epsilon_o\mu_o)\bar{v}_o$$

$\bar{v}_o$ is the translatory velocity of the medium,

ϵ, μ are the permittivity and the permeability of the material medium in its rest frame, and

ϵ_o, μ_o are the permittivity and permeability of free space.

From (A-4) the magnetic vector potential $\bar{A}$ may be defined:

$$\bar{B} = \nabla \times \bar{A}. \tag{A-7}$$

Using (A-7) in (A-1), the electric scalar potential Φ may be defined:

$$\bar{E} = -\nabla\Phi - \frac{\partial \bar{A}}{\partial t}. \tag{A-8}$$

Substituting (A-7) and (A-8) in (A-5) and rearranging, one has:

$$\bar{H} = \frac{1}{\mu}\left[\nabla \times \bar{A} - \bar{\Lambda} \times \left(\nabla\Phi + \frac{\partial \bar{A}}{\partial t}\right)\right] \tag{A-9}$$

Using (A-6), (A-8), and (A-9) in (A-2) and (A-3), and rearranging, one obtains:

$$\left(\nabla - \bar{\Lambda}\frac{\partial}{\partial t}\right) \times \left(\nabla - \bar{\Lambda}\frac{\partial}{\partial t}\right) \times \bar{A} - \nabla \times (\bar{\Lambda} \times \nabla\Phi) =$$

$$= \mu\bar{J} - \nabla\left(\mu\epsilon\frac{\partial\Phi}{\partial t}\right) - \mu\epsilon\frac{\partial^2 \bar{A}}{\partial t^2}\ ; \tag{A-10}$$

$$\nabla^2\Phi + \nabla \cdot \frac{\partial \bar{A}}{\partial t} + \frac{1}{\mu\epsilon}(\nabla \times \nabla \times \bar{A}) \cdot \bar{\Lambda} = -\frac{\rho}{\epsilon} \tag{A-11}$$

where Λ^2 is neglected for the non-relativistic approximation. Multiplying (A-11) by $\bar{\Lambda}$, one has:

$$\bar{\Lambda}\nabla^2\Phi = -\frac{\rho}{\epsilon}\bar{\Lambda} - \bar{\Lambda}\,\nabla \cdot \frac{\partial \bar{A}}{\partial t}. \tag{A-12}$$

Expanding (A-10) and using (A-12), one obtains:

$$\left[\nabla^2 - \mu\epsilon\frac{\partial^2}{\partial t^2} - 2\bar{\Lambda} \cdot \nabla\frac{\partial}{\partial t}\right]\bar{A} -$$

$$-\nabla\left[\nabla \cdot \bar{A} - \bar{\Lambda} \cdot \frac{\partial \bar{A}}{\partial t} + \bar{\Lambda} \cdot \nabla\Phi + \mu\epsilon\frac{\partial\Phi}{\partial t}\right] = -\mu\bar{J} + \frac{\rho\bar{\Lambda}}{\epsilon} \tag{A-13}$$

where the identity $(\bar{\Lambda} \cdot \nabla)\nabla\Phi = \nabla(\bar{\Lambda} \cdot \nabla\Phi)$ is used. The wave equation for $\bar{A}$ may be obtained by imposing the condition:

$$\nabla \cdot \bar{A} - \bar{\Lambda} \cdot \frac{\partial \bar{A}}{\partial t} + \bar{\Lambda} \cdot \nabla\Phi + \mu\epsilon\frac{\partial\Phi}{\partial t} = 0. \tag{A-14}$$

Equation (A-14) is the Lorentz condition for the case under consideration. With this condition, the vector potential wave equation (A-13) becomes:

$$\left[\nabla^2 - \mu\epsilon\frac{\partial^2}{\partial t^2} - 2\overline{\Lambda}\cdot\nabla\frac{\partial}{\partial t}\right]\overline{A} = -\mu\overline{J} + \frac{1}{\epsilon}\rho\overline{\Lambda} . \tag{A-15}$$

Taking the scalar product of (A-15) with $\overline{\Lambda}$ and using it together with the Lorentz condition (A-14) in (A-11), one obtains the scalar potential wave equation:

$$\left[\nabla^2 - \mu\epsilon\frac{\partial^2}{\partial t^2} - 2\overline{\Lambda}\cdot\nabla\frac{\partial}{\partial t}\right]\Phi = -\frac{1}{\epsilon}(\rho + \overline{\Lambda}\cdot\overline{J}). \tag{A-16}$$

Equations (A-15) and (A-16) agree with Papas [1965] for the particular case of non-relativistic velocities. Similar equations may be obtained for the electric vector potential and the magnetic scalar potential, for a homogeneous, isotropic, moving medium with magnetic current sources.

APPENDIX B

The q Transformation

The wave vector transformation is given by (2-25), as follows:

$$\bar{k}' = \bar{k} - \gamma \frac{\omega}{c^2} \bar{v}_o + (\gamma - 1) \frac{\bar{k} \cdot \bar{v}_o}{v_o^2} \bar{v}_o \tag{B-1}$$

Then, for the wave in the x-z plane, transmitted in the plasma, moving in the $\hat{z}$ direction, one has:

$$\bar{k}_t' = k_{tx}\hat{x} + k_{tz}\hat{z} - \gamma \frac{\omega_t}{c^2} v_o\hat{z} + \gamma k_{tz}\hat{z} - k_{tz}\hat{z}, \tag{B-2}$$

so that one obtains from (B-2):

$$k_{tx}' = k_{tx};\ k_{tz}' = \gamma\left(k_{tz} - \frac{\omega_t}{c^2} v_o\right) \tag{B-3}$$

From the latter expression in (B-3) one obtains:

$$k_{tz}' = k_o'q' = \gamma\left[k_{to}q - \frac{\omega_t}{c^2} v_o\right] \tag{B-4}$$

where we define from (2-26):

$$k_o' = \omega'/c = \gamma\omega_t(1 - q\beta)/c$$

$$k_{to} = \omega_t/c;\ \beta = v_o/c$$

hence (B-4) becomes:

$$q' = (q - \beta)/(1 - q\beta) \tag{B-5}$$

from which it follows that:

$$q = (q' + \beta)/(1 + q'\beta). \tag{B-6}$$

APPENDIX C

The Resonance Frequency Transformation

Taking the counted number of natural oscillations of a bounded electron in the rest frame of reference S' during an interval of time $\nabla t'$, which corresponds to the same counted number of natural oscillations of the bounded electron in the laboratory frame of reference S during a corresponding interval of time Δt, the corresponding resonance circular frequencies ω_0' and ω_0 will be defined as follows:

$$\omega_0' = 2\pi \frac{\text{No. Oscillations}}{\Delta t'}; \quad \omega_0 = 2\pi \frac{\text{No. Oscillations}}{\Delta t} \qquad \text{(C-1)}$$

In accordance with Einstein's relativistic time dilatation effect, the shortest time interval is shown by the proper clock in the proper system [Pauli, 1958], as follows:

$$\Delta t' = \Delta t/\gamma = \Delta t\sqrt{1 - (v_0/c)^2} \qquad \text{(C-2)}$$

Substituting (C-2) in (C-1), one obtains [Chawla and Unz, 1967b]:

$$\omega_0' = \gamma\omega_0 \qquad \text{(C-3)}$$

where (C-3) represents the resonance frequency transformation. Equation (C-3) agrees with the experimental work on the Mossbauer effect [Elliott, 1966], and may be used to derive the corresponding refractive index equations for bounded magneto-plasma [Chawla and Unz, 1967b].

REFERENCES

Allis, W. P., S. J. Buchsbaum, and A. Bers (1963), Waves in Anisotropic Plasmas (M.I.T. Press).

Aris, H. G. (1962), Vectors, Tensors and the Basic Equations of Fluid Mechanics (Prentice-Hall).

Bailey, V. A. (1948), Plane waves in an ionized gas with static electric and magnetic fields present, Aust. J. Sci. Res., A -1, 351-359.

Bailey, V. A. (1950), The growth of circularly polarized waves in the sun's atmosphere and their escape into space, Phys. Rev., 78, 428-443.

Bailey, V. A. (1951), The relativistic theory of electro-magneto-ionic waves, Phys. Rev., 83, 439-454.

Beck, F. (1953), Der energie-impulstensor in der phanomenologischen theorie der supraleitung, Zeit. Physik, 134, 334-345.

Becker, R., and F. Sauter (1964), Electromagnetic Fields and Interactions (Blaisdell).

Bell, T., and R. A. Helliwell (1960), Traveling wave amplification in the ionosphere, Proc. Symp. on Physical Processes in the Sun-Earth Environment, D.R.B., Canada, 215-222.

Bell, T., R. Smith, N. B. Brice, and H. Unz (1963), The magneto-ionic theory for a drifting plasma, IEEE Trans., AP-11, 194-195.

Booker, H. G. (1936), Oblique propagation of electromagnetic waves in a slowly varying non-isotropic medium, Proc. Roy. Soc., A-155, 235.

Booker, H. G. (1939), The propagation of wave packets incident obliquely on a stratified doubly refracting ionosphere, Phil Trans., A-237, 441.

Booker, H. G. (1949), The application of the magneto-ionic theory to radio waves incident obliquely upon a horizontally stratified ionosphere, J. Geophys. Res., 54, 243.

Brandstatter, J. J. (1963), Waves, Rays and Radiation in Plasma Media (McGraw Hill).

Briggs, R. J. (1964), Electron Stream Interactions with Plasmas (M.I.T. Press).

Budden, K. G. (1961), Radio Waves in the Ionosphere (Cambridge Univ. Press).

Chawla, B. R., S. S. Rao, and H. Unz (1966), Wave propagation in homogeneous gyrotropic compressible drifting plasma, J. Appl. Phys., 37, 3563-3566.

Chawla, B. R., and H. Unz (1966a), Radiation in a moving anisotropic plasma, IEEE Proc., 54, 1103-1105.

Chawla, B. R., and H. Unz (1966b), Relativistic magneto-ionic theory for drifting compressible plasma in the longitudinal direction, IEEE Proc., 54, 1214-1215.

Chawla, B. R., and H. Unz (1966c), Potential equations in homogeneous isotropic moving media, IEEE Proc., 54, 397.

Chawla, B. R., and H. Unz (1967a), Normal incidence on semi-infinite longitudinally drifting magneto-plasma: the relativistic solution, IEEE Trans., AP-15, 324-326.

Chawla, B. R., and H. Unz (1967b), Wave propagation in a moving plasma, Electronics Letters, 3, 244-246.

Chawla, B. R., and H. Unz (1968), Reflection and transmission of normally incident waves by a semi-infinite longitudinally drifting magneto-plasma, Il Nuovo Cimento, 57-B, 399-418.

Chawla, B. R., and H. Unz (1969a), Wave propagation in relativistically moving warm magneto-plasma, IEEE Trans., AP-17, 822-823.

Chawla, B. R., and H. Unz (1969b), Reflection and transmission of electromagnetic waves normally incident on a plasma slab moving uniformly along a magneto-static field, IEEE Trans., AP-17, 771-777.

Chawla, B. R., and H. Unz (1969c), Propagation in a two stream magneto-plasma, IEEE Trans., AP-17, 384-385.

Chen, H. C., and D. K. Cheng (1966), Constitutive relations for a moving anisotropic medium, IEEE Proc., 54, 62-63.

Clemmow, P. C. (1962), Wave amplification in a plasma stream in a medium of high refractive index, Proc. Phys. Soc. (London), 80, 1322-1332.

Einstein, A., H. A. Lorentz, H. Weyl, and H. Minkowski (1923), The Principle of Relativity (Dover).

Elliott, R. E. (1966), Electromagnetics (McGraw Hill).

Epstein, M., and H. Unz (1963), Comments on two papers dealing with electromagnetic wave propagation in moving plasmas, IEEE Trans., AP-11, 193-194.

Fainberg, I. B., and V. S. Tkalich (1959), On the reflection of an electromagnetic wave from the plasma moving through a dielectric in a constant magnetic field, Sov. Phys. Tech. Phys., 4, 438-443.

Fano, R. M., L. J. Chu, and R. B. Adler (1960), Electromagnetic Fields, Energy and Forces (Wiley).

Frank, I. (1943), Doppler effect in a refractive medium, J. Phys. USSR, 7, 49-67.

Getmantsev, G. G., and V. O. Rapoport (1960), On the build-up of electromagnetic waves in a plasma moving in a non-dispersive dielectric in the presence of a constant magnetic field, Sov. Phys. JETP, 11, 871-875.

Ginzburg, V. L. (1963), The Propagation of Electromagnetic Waves in Plasmas (Addison-Wesley).

Holt, E. H., and R. E. Haskell (1965), Plasma Dynamics (Macmillan).

Koyama, M., B. R. Chawla, and H. Unz (1967), On the polarization and the convection current models, IEEE Proc., 55, 579-581.

Lampert, M. A. (1956), Incidence of an electromagnetic wave on a 'Cerenkov electron gas,' Phys. Rev., 102, 299-304.

Landau, L. D., and E. M. Lifshitz (1966), Electrodynamics of Continuous Media (Addison-Wesley).

Landecker, K. (1952), Possibility of frequency multiplication and wave amplification by means of some relativistic effects, Phys. Rev., 86, 852-855.

Lee, K. S. H., and C. H. Papas (1964), Electromagnetic radiation in the presence of moving simple media, J. Math. Phys., 5, 1668-1672.

Lee, K. S. H., and C. H. Papas (1965), Antenna radiation in a moving dispersive medium, IEEE Trans., AP-13, 799-804.

Lee, S. W., and Y. T. Lo (1966), Radiation in a moving anisotropic medium, Radio Science, 1, 313-324.

Lorentz, H. A. (1952), The Theory of Electrons (Dover).

Meyers, N. H. (1958), A Poynting theorem for moving bodies and the relativistic mechanics of extended objects, J. Franklin Inst., 266, 439-464.

Møller, C. (1952), The Theory of Relativity (Oxford Univ. Press).

Oster, L. (1960), Linearized theory of plasma oscillations, Rev. Mod. Phys., 32, 141-169.

Panofsky, W. K. H., and M. Phillips (1962), Classical Electricity and Magnetism (Addison-Wesley).

Papas, C. H. (1965), Theory of Electromagnetic Wave Propagation (McGraw-Hill).

Pauli, W. (1958), Theory of Relativity (Pergamon Press).

Ratcliffe, J. A. (1959), The Magneto-Ionic Theory and its Applications to the Ionosphere (Cambridge Univ. Press).

Scarf, F. L. (1961), Wave propagation in a moving plasma, Am. J. Phys., 29, 101-107.

Seshadri, S. R. (1964), Wave propagation in a compressible ionosphere, part I, Radio Sci. J. Res. NBS, 68D, 1285-1295.

Sidhu, D. P., and H. Unz (1967), The magneto-ionic theory for drifting plasma--the Whistler mode, Kansas Acad. Sci. Trans., 70, 432-450.

Sokolnikoff, I. S. (1951), Tensor Calculus (Wiley).

Sommerfeld, A. (1964a), Electrodynamics (Academic Press).

Sommerfeld, A. (1964b), Optics (Academic Press).

Stratton, J. A. (1941), Electromagnetic Theory (McGraw-Hill).

Tai, C. T. (1964a), A study of electrodynamics of moving media, IEEE Proc., 52, 685-689.

Tai, C. T. (1964b), Two scattering problems involving moving media, The Antenna Lab., Ohio State University, Report 1691-7, Columbus, Ohio.

Tai, C. T. (1965a), Electrodynamics of moving anisotropic media: The first order theory, Radio Sci. J. Res. NBS, 69D, 401-405.

Tai, C. T. (1965b), The dyadic Green's function for a moving isotropic medium, IEEE Trans., AP-13, 322-323.

Tai, C. T. (1965c), Huygen's principle in a moving, isotropic, homogeneous and linear medium, App. Optics, 4, 1347-1349.

Tai, C. T. (1967), Present views on electrodynamics of moving media, Radio Science, 2, 245-248.

Tang, C. L., and J. Meixner (1961), Relativistic theory of the propagation of plane electromagnetic waves, Phys. Fluids, 4, 148-154.

Twiss, R. Q. (1951), On Bailey's theory of amplified circularly polarized waves in an ionized medium Phys. Rev., 84, 448-457.

Unz, H. (1961), On vertical drift velocities of F-2 layer, J. Atmos. Terr. Phys., 21, 237-242.

Unz, H. (1962), The magneto-ionic theory for drifting plasma, IRE Trans., AP-10, 459-464.

Unz, H. (1963), Electromagnetic radiation in drifting Tellegen anisotropic medium, IEEE Trans., AP-11 573-578.

Unz, H. (1965a), Drifting plasma magneto-ionic theory of oblique incidence, IEEE Trans., AP-13, 595-600.

Unz, H. (1965b), Oblique incidence on plane boundary between two general gyrotropic plasma media, J. Math. Phys., 6, 1813-1821.

Unz, H. (1966a), Relativistic magneto-ionic theory for drifting plasma in longitudinal direction, Phys. Rev., 146, 92-95.

Unz, H. (1966b), Oblique wave propagation in a compressible general magneto-plasma, Appl. Sci. Res., 16, 105-120.

Unz, H. (1966c), On the derivation of Booker's quartic from Appleton-Hartree equation, IEEE Proc., 54, 304.

Unz, H. (1966d), Wave propagation in drifting isotropic warm plasma, Radio Science, 1, 325-338.

Unz, H. (1967a), Normal incidence on semi-infinite longitudinally drifting magneto-plasma: The non-relativistic solution, IEEE Trans., MTT-15, 432-433.

Unz, H. (1967b), Oblique electromagnetic radiation incident on a semi-infinite warm general magnetoplasma, Il Nuovo Cimento, B-50, 207-223.

Unz, H. (1968), Relativistic magneto-ionic theory for drifting plasma, Radio Science, 3, 295-298.

Yeh, C. (1965), Reflection and transmission of electromagnetic waves by a moving dielectric medium, J. Appl. Phys., 36, 3513-3517.

Yeh, C. (1966), Reflection and transmission of electromagnetic waves by a moving plasma medium, J. Appl. Phys., 37, 3079-3082.

Yeh, C., and K. F. Casey (1966), Reflection and transmission of electromagnetic waves by a moving dielectric slab, Phys. Rev., 144, 665-669.

BIBLIOGRAPHY

BOOKS

ARTICLES

BOOKS

A. TENSOR ANALYSIS

1. Aris, R. (1962), Vectors, Tensors and the Basic Equations of Fluid Mechanics (Prentice-Hall).
2. Brand, L. (1947), Vector and Tensor Analysis (Wiley).
3. Holt, E. H., and R. E. Haskell (1965), Foundations of Plasma Dynamics (MacMillan).
4. Jeffreys, H. (1957), Cartesian Tensors (Cambridge Univ. Press).
5. Krishnamurty, K. (1967), Vector Analysis and Cartesian Tensors (Holden-Day).
6. Kyrala, A. (1967), Theoretical Physics, Applications of Vectors, Matrices, Tensors and Quaternions (Saunders).
7. Lass, H. (1950), Vector and Tensor Analysis (McGraw-Hill).
8. McConnell, A. J. (1957), Applications of Tensor Analysis (Dover).
9. Moon, P., and D. E. Spencer (1965), Vectors (Van Nostrand).
10. Prager, W. (1961), Introduction to Mechanics of Continua (Ginn).
11. Sokolnikoff, I. S. (1951), Tensor Calculus (Wiley).
12. Spain, B. (1953), Tensor Calculus (Central).
13. Spiegel, M. R. (1959), Vector Analysis and Introduction to Tensor Analysis (Schaum-McGraw-Hill)
14. Springer, C. E. (1962), Tensor and Vector Analysis (Ronald Press).
15. Stigant, S. A. (1959), The Elements of Determinants, Matrices, and Tensors for Engineers (MacDonald).
16. Stigant, S. A. (1964), Applied Tensor Analysis for Electrical Students (MacDonald).

B. RELATIVITY AND ELECTROMAGNETIC THEORY

17. Aharoni, J. (1959), The Special Theory of Relativity (Oxford Univ. Press).
18. Anderson, N. (1968), The Electromagnetic Field (Plenum Press).
19. Becker, R., and F. Sauter (1964), Electromagnetic Fields and Interactions, vol. I: Electromagnetic Theory and Relativity (Blaisdell).
20. Bergman, P. G. (1942), An Introduction to the Theory of Relativity (Prentice-Hall).
21. Bohm, D. (1965), The Special Theory of Relativity (Benjamin).
22. Born, M. (1962), Einstein's Theory of Relativity (Dov

23. Born, M., and E. Wolf (1964), Principles of Optics (Pergamon Press).
24. Brillouin, L. (1960), Wave Propagation and Group Velocity (Academic Press).
25. Clemmow, P. C. (1966), The Plane Wave Spectrum Representation of Electromagnetic Fields (Pergamon Press).
26. Corson, D., and P. Lorrain (1962), Introduction to Electromagnetic Fields and Waves (Freeman).
27. Cowan, E. W. (1968), Basic Electromagnetism (Academic Press).
28. Cullwick, E. G. (1959), Electromagnetism and Relativity (Wiley).
29. Eddington, A. S. (1930), The Mathematical Theory of Relativity (Cambridge Univ. Press).
30. Einstein, A. (1956), The Meaning of Relativity (Princeton Univ. Press).
31. Einstein, A. (1961), Relativity, The Special and General Theory (Crown).
32. Einstein, A., H. A. Lorentz, H. Minkowski, and H. Weyl (1952), The Principle of Relativity (Dover).
33. Elliott, R. S. (1966), Electromagnetics (McGraw-Hill).
34. Fano, R. M., L. J. Chu, and R. B. Adler (1960), Electromagnetic Fields, Energy and Forces (Wiley).
35. Fock, V. (1959), The Theory of Space, Time and Gravitation (Pergamon).
36. Gill, T. P. (1965), The Doppler Effect (Logos Press).
37. Guggenheim, A. (1967), Elements and Formulae of Special Relativity (Pergamon).
38. Hagedorn, R. (1964), Relativistic Kinematics (Benjamin).
39. Hallen, E. (1962), Electromagnetic Theory (Wiley).
40. Heading, J. (1964), Electromagnetic Theory and Special Relativity (Univ. Tutorial Press, London).
41. Helliwell, T. M. (1966), Introduction to Special Relativity (Allyn and Bacon).
42. Jackson, J. D. (1962), Classical Electrodynamics (Wiley).
43. Javid, M., and P. M. Brown (1963), Field Analysis and Electromagnetics (McGraw-Hill).
44. Jones, D. S. (1964), The Theory of Electromagnetism (MacMillan).
45. Jordan, E. C., and K. G. Balmain (1968), Electromagnetic Waves and Radiation Systems (Prentice-Hall).

46. Kacser, C. (1967), Introduction to the Special Theory of Relativity (Prentice-Hall).
47. King, R. W. P. (1945), Electromagnetic Engineering (McGraw-Hill).
48. King, R. W. P. (1963), Fundamental Electromagnetic Theory (Dover).
49. Landau, L. D., and E. M. Lifshitz (1969), The Classical Theory of Fields (Addison-Wesley).
50. Landau, L. D., and E. M. Lifshitz (1960), Electrodynamics of Continuous Media (Addison-Wesley).
51. Lawden, D. F. (1967), An Introduction to Tensor Calculus and Relativity (Melhuen).
52. Lichnerowicz, A. (1967), Relativistic Hydrodynamics and Magnetohydrodynamics (Benjamin).
53. Mermin, N. D. (1968), Space and Time in Special Relativity (McGraw-Hill).
54. Møller, C. (1952), The Theory of Relativity (Oxford Univ. Press).
55. Morse, P. M., and H. Feshbach (1953), Methods of Theoretical Physics, vols. I, II (McGraw-Hill).
56. Morse, P. M., and K. U. Ingard (1968), Theoretical Acoustics (McGraw-Hill).
57. Naiwark, M. A. (1964), Linear Representations of the Lorentz Group (Pergamon Press).
58. Ney, E. P. (1962), Electromagnetism and Relativity (Harper and Row).
59. Panofsky, W. K. H., and M. Phillips (1962), Classical Electricity and Magnetism (Addison-Wesley).
60. Papas, C. H. (1965), Theory of Electromagnetic Wave Propagation (McGraw-Hill).
61. Pauli, W. (1958), Theory of Relativity (Pergamon Press).
62. Penfield, P., and H. A. Haus (1967), Electrodynamics of Moving Media (M.I.T. Press).
63. Plonsey, R., and R. E. Collin (1961), Principles and Applications of Electromagnetic Fields (McGraw-Hill).
64. Post, E. J. (1962), Formal Structure of Electromagnetics (North-Holland).
65. Prokhovnik, S. J. (1967), The Logic of Special Relativity (University Press).
66. Ramo, S., J. R. Whinnery, and T. VanDuzer (1965), Fields and Waves in Communication Electronics (Wiley).
67. Reitz, J. R., and F. J. Milford (1960), Foundations of Electromagnetic Theory (Addison-Wesley).
68. Resnick, R. (1968), Introduction to Special Relativity (Wiley).
69. Rindler, W. (1960), Special Relativity (Interscience

70. Rosser, W. G. V. (1968), Classical Electromagnetism via Relativity (Plenum Press).
71. Rosser, W. G. V. (1968), Introductory Relativity (Plenum Press).
72. Sanders, J. H. (1965), Velocity of Light (Pergamon Press).
73. Schwartz, H. M. (1968), Introduction to Special Relativity (McGraw-Hill).
74. Shadowitz, A. (1968), Special Relativity (Saunders).
75. Silverstein, L. (1914), The Theory of Relativity (MacMillan).
76. Smith, J. H. (1965), Introduction to Special Relativity (Benjamin).
77. Smythe, W. R. (1950), Static and Dynamic Electricity (McGraw-Hill).
78. Sommerfeld, A. (1964), Electrodynamics (Academic Press).
79. Sommerfeld, A. (1964), Optics (Academic Press).
80. Stratton, J. A. (1941), Electromagnetic Theory (McGraw-Hill).
81. Synge, J. L. (1958), Relativity, The Special Theory (North-Holland).
82. Terletskii, Ya. P. (1968), Paradoxes in the Theory of Relativity (Plenum Press).
83. Tolman, R. C. (1917), The Theory of Relativity of Motion (Univ. of California Press).
84. Tolman, R. C. (1934), Relativity, Thermodynamics and Cosmology (Oxford Univ. Press).
85. Tonnelat, M. A. (1966), Einstein's Unified Field Theory (Gordon and Breach).
86. Tralli, N. (1963), Classical Electromagnetic Theory (McGraw-Hill).
87. Tyras, G. (1969), Radiation and Propagation of Electromagnetic Waves (Academic Press).
88. Van Bladel, J. (1964), Electromagnetic Fields (McGraw-Hill).
89. Whittaker, E. (1953), A History of the Theories of Aether and Electricity: vol. I, Classical Theories, vol. II, The Modern Theories 1900-1926 (Nelson, London).
90. Yilmaz, H. (1965), Theory of Relativity and the Principles of Modern Physics (Blaisdell).

C. MICROWAVE TUBES

91. Angelakos, D. J., and T. E. Everhart (1968), Microwave Communications (McGraw-Hill).
92. Atwater, H. A. (1962), Introduction to Microwave Theory (McGraw-Hill).
93. Beck, A. H. W. (1958), Space Charge Waves (Pergamon Press).

94. Birdsall, C. K., and W. B. Bridges (1966), Electron Dynamics of Diode Regions (Academic Press).
95. Brillouin, L. (1953), Wave Propagation in Periodic Structures (Dover).
96. Chodorow, M., and C. Susskind (1964), Fundamentals of Microwave Electronics (McGraw-Hill).
97. Collin. R. E. (1966), Foundations of Microwave Engineering (McGraw-Hill).
98. Harman, W. W. (1953), Fundamentals of Electronic Motion (McGraw-Hill).
99. Harvey, A. F. (1963), Microwave Engineering (Academic Press).
100. Hutter, R. G. E. (1960), Beam and Wave Electronics in Microwave Tubes (Van Nostrand).
101. Ishii, T. K. (1966), Microwave Engineering (Ronald Press).
102. Kirstein, P. T. (1967), Space-Charge Flow (McGraw-Hill).
103. Kleen, W. J. (1958), Electronics of Microwave Tubes (Academic Press).
104. Pierce, J. R. (1949), Theory and Design of Electron Beams (Van Nostrand).
105. Pierce, J. R. (1950), Traveling Wave Tubes (Van Nostrand).
106. Reintjes, J. F., and G. T. Coate (1952), Principles of Radar (McGraw-Hill).
107. Shevchik, V. N., G. N. Shvedov, and A. V. Soboleva (1966), Wave and Oscillatory Phenomena in Electron Beams at Microwave Frequencies (Pergamon Press).
108. Skolnik, M. I. (1962), Radar Systems (McGraw-Hill).
109. Slater, J. C. (1954), Microwave Electronics (Van Nostrand).
110. Spangenberg, K. R. (1948), Vacuum Tubes (McGraw-Hill).
111. Spangenberg, K. R. (1957), Fundamentals of Electron Devices (McGraw-Hill).

D. THE IONOSPHERE

112. Al'pert, Ya. L. (1963), Radio Wave Propagation and the Ionosphere (Plenum Press).
113. Beynon, W. J. G. (Editor) (1962), Monograph on Ionospheric Radio (Elsevier).
114. Boyd, R. L. F., and M. J. Seaton (Editor) (1954), Rocket Exploration of the Upper Atmosphere (Pergamon Press).
115. Bremmer, H. (1949), Terrestrial Radio Waves (Elsevier).
116. Brown, G. M. (Editor) (1965), Progress in Radio Science 1960-1963, vol. III: The Ionosphere (Elsevier).

117. Budden, K. G. (1961), Radio Waves in the Ionosphere (Cambridge Univ. Press).
118. Budden, K. G. (1961), The Waveguide Mode Theory of Wave Propagation (Prentice-Hall).
119. Budden, K. G. (1964), Lectures on Magneto-Ionic Theory (Gordon and Breach).
120. Craig, R. A. (1965), The Upper Atmosphere (Academic Press).
121. Davies, K. (1966), Ionospheric Radio Propagation (Dover).
122. Folkestad, K. (Editor) (1968), Ionospheric Radio Communications (Plenum Press).
123. Helliwell, R. A. (1965), Whistlers and Related Ionospheric Phenomena (Stanford Univ. Press).
124. Hines, C. O., I. Paghis, T. R. Hartz, and J. A. Fejer (1965), Physics of the Earth Upper Atmosphere (Prentice-Hall).
125. Kelso, J. M. (1964), Radio Ray Propagation in the Ionosphere (McGraw-Hill).
126. Kerr, D. K. (1948), Propagation of Short Radio Waves, MIT Radiation Lab. Series, vol. 13 (McGraw-Hill).
127. Maeda, K. I., and S. Silver (Editors) (1965), Progress in Radio Science 1960-1963: vol. 8: Space Radio Science (Elsevier).
128. Maehlum, B. (Editor) (1962), Electron Density Profiles in the Ionosphere and Exosphere (Pergamon Press).
129. Manning, L. A. (1962), Bibliography of the Ionosphere (Stanford Univ. Press).
130. Massey, H. S. W., and R. L. F. Boyd (1958), The Upper Atmosphere (Hutchinson, London).
131. Mitra, S. K. (1952), The Upper Atmosphere (The Asiatic Society, Calcutta, India).
132. Nupen, W. (1960), Bibliography on Ionospheric Propagation of Radio Waves 1923-1960, National Bureau of Standards, U.S. Department of Commerce, Technical Note No. 84, Boulder, Colorado.
133. Physical Soc. (1955), The Physics of the Ionosphere, Report of Conference (The Physical Society).
134. Ratcliffe, J. A. (1959), The Magneto-Ionic Theory and Its Applications To The Ionosphere (Cambridge Univ. Press).
135. Ratcliffe, J. A. (Editor) (1960), Physics of the Upper Atmosphere (Academic Press).
136. Rawer, K. (1952), The Ionosphere (Ungar).
137. Van Allen, J. A. (1956), Scientific Uses of Earth Satellites (Univ. Michigan Press).

138. Wait, J. R. (1962), Electromagnetic Waves in Strati fied Media (Pergamon Press).
139. Whitten, R. C., and I. G. Poppoff (1965), Physics of the Lower Ionosphere (Prentice-Hall).

E. PLASMA PHYSICS - TEXTBOOKS

140. Allis, W. P., S. J. Buchsbaum, and A. Bers (1963), Waves in Anisotropic Plasmas (M.I.T. Press
141. Bekefi, G. (1966), Radiation Processes in Plasmas (Wiley).
142. Brandstatter, J. J. (1963), An Introduction to Wave Rays and Radiation in Plasma Media (McGraw Hill).
143. Campbel, A. B. (1963), Plasma Physics and Magneto Fluid-Mechanics (McGraw-Hill).
144. Chandrasekhar, S. (1960), Plasma Physics (Chicago Univ. Press).
145. Clemmow, P. C., and J. P. Dougherty (1969), Electro dynamics of Particles and Plasmas (Addison Wesley).
146. Delcroix, J. L. (1965), Plasma Physics (Wiley).
147. Delcroix, J. (1969), Plasma Physics, vol. 2: Weakly Ionized Gases (Wiley).
148. Denisse, J. F., and J. L. Delcroix (1963), Plasma Waves (Interscience).
149. Ferraro, V. C. A., and C. Plumpton (1966), An Intro duction to Magneto-Fluid Mechanics (Oxford Univ. Press).
150. Gartenhaus, S. (1964), Elements of Plasma Physics (Holt, Rinehart and Winston).
151. Ginzburg, V. L. (1964), The Propagation of Electro-magnetic Waves in Plasmas (Addison-Wesley)
152. Green, H. S., and R. B Leipnik (1966), Magneto-hydrodynamics and Plasma Physics (Academic Press).
153. Hellund, E. (1961), The Plasma State (Reinhold).
154. Holt, E. H., and R. E. Haskell (1965), Foundations of Plasma Dynamics (MacMillan).
155. Hughes, W. F., and F. J. Young (1966), Electro-magnetodynamics of Fluids (Wiley).
156. Jancel, R., and Th. Kahan (1966), Electrodynamics of Plasmas (Wiley).
157. Kampen, Van N. G., and B. U. Felderhof (1967), Theoretical Methods in Plasma Physics (North-Holland).
158. Linhart, J. G. (1961), Plasma Physics (North-Holland).
159. Longmire, C. L. (1963), Elementary Plasma Physics (Interscience).
160. Montgomery, D. C., and D. A. Tidman (1964), Plasma Kinetic Theory (McGraw-Hill).

161. Pai, S. I. (1962), Magnetogasdynamics and Plasma Dynamics (Prentice-Hall).
162. Robert, P. H. (1967), An Introduction to Magnetohydrodynamics (Elsevier).
163. Rose, D. J., and M. Clark (1961), Plasmas and Controlled Fusion (Wiley).
164. Shkarofsky, I. P., T. W. Johnston, and M. P. Bachynski (1966), The Particle Kinetics of Plasmas (Addison-Wesley).
165. Stix, T. H. (1962), The Theory of Plasma Waves (McGraw-Hill).
166. Tanenbaum, B. S. (1967), Plasma Physics (McGraw-Hill).
167. Thompson, W. B. (1962), An Introduction to Plasma Physics (Addison-Wesley).
168. Uman, M. A. (1964), Introduction to Plasma Physics (McGraw-Hill).
169. Wait, J. R. (1968), Electromagnetics and Plasmas (Holt, Rinehart and Winston).
170. Wu, T. Y. (1966), Kinetic Equations of Gases and Plasmas (Addison-Wesley).

F. PLASMA PHYSICS - RELATED TEXTS

171. Adams, R. N., and E. D. Denman (1967), Wave Propagation in Turbulent Media, (Elsevier).
172. Akhiezer, A. I., et al. (1967), Collective Oscillations in a Plasma (M.I.T. Press).
173. Alfven, H., and C. G. Falthammer (1963), Cosmical Electrodynamics (Oxford Univ. Press).
174. Allis, W. P. (1956), Motion of Ions and Electrons, Handbuch der Physik, S. Flugge (Editor) (Springer-Verlag, 21, 383-444.
175. Anderson, J. E. (1963), Magnetohydrodynamic Shock Waves (M.I.T. Press).
176. Aukland, M. F. (1963), Plasma Physics and Magnetohydrodynamics, A Report Bibliography, Defense Documentation Center AD-271-170 and AD-405-732.
177. Artsimovich, G., R. S. Pease, and A. C. Kolb (1964), Controlled Thermonuclear Reactions (Gordon and Breach).
178. Balescu, R. (1963), Statistical Mechanics of Charged Particles (Interscience).
179. Bird, R. B., W. E. Stewart, and E. N. Lightfoot (1960), Transport Phenomena (Wiley).
180. Brekhovshikh, L. M. (1960), Waves in Layered Media (Academic Press).
181. Briggs, R. J. (1964), Electron-Stream Interaction with Plasmas (M.I.T. Press).

182. Brown, S. C. (1967), Basic Data of Plasma Physics (M.I.T. Press).
183. Chapman, S. (1964), Solar Plasma, Geomagnetism and Aurora (Gordon and Breach).
184. Chapman, S., and T. G. Cowling (1964), Mathematical Theory of Nonuniform Gases (Cambridge Univ. Press).
185. Chernov, L. A. (1960), Wave Propagation in a Random Medium (McGraw-Hill).
186. Cowling, T. S. (1957), Magnetohydrodynamics (Interscience).
187. Delcroix, J. L. (1960), Introduction to the Theory of Ionized Gases (Interscience).
188. Dungey, J. W. (1958), Cosmic Electrodynamics (Cambridge Univ. Press).
189. Fried, B. D., and S. D. Conte (1961), The Plasma Dispersion Function (Academic Press).
190. Goldstein, S. (1960), Lectures on Fluid Mechanics (Interscience).
191. Griem, H. R. (1964), Plasma Spectroscopy (McGraw-Hill).
192. Heald, M. A., and C. B. Wharton (1965), Plasma Diagnostics with Microwaves (Wiley).
193. Hill, T. L. (1956), Statistical Mechanics (McGraw-Hill).
194. Hill, T. L. (1960), Introduction to Statistical Thermodynamics (Addison-Wesley).
195. Huang, K. (1963), Statistical Mechanics (Wiley).
196. Huddlestone, R. H., and S. L. Leonard (1965), Plasma Diagnostic Techniques (Academic Press).
197. Hughes, W. F., and E. W. Gaylord (1964), Basic Equations of Engineering Science (Schaum).
198. Jeans, J. (1925), The Dynamic Theory of Gases (Dover).
199. Jelley, J. V. (1958), Cerenkov Radiation and its Applications (Pergamon Press).
200. Kadomtsev, B. B. (1965), Plasma Turbulence (Academic Press).
201. Kalikhman, L. E. (1967), Elements of Magnetogasdynamics (Saunders).
202. Klimontovich, Yu. L. (1967), The Statistical Theory of Non-Equilibrium Processes in a Plasma (M.I.T. Press).
203. Landau, L. D., and E. M. Lifshitz (1955), Statistical Physics (Pergamon Press).
204. Lehnert, B. (1964), Dynamics of Charged Particles (Interscience).
205. Llewellyn, J. F. (1966), The Glow Discharge and an Introduction to Plasma Physics (Wiley).
206. Lorentz, H. A. (1952), The Theory of Electrons (Dove

207. McDaniel, E. W. (1964), Collision Phenomena in Ionized Gases (Wiley).
208. Melcher, J. R. (1963), Field-Coupled Waves (M.I.T. Press).
209. Present, R. D. (1958), Kinetic Theory of Gases (McGraw-Hill).
210. Ramer, J. D. (1959), Bibliography on Plasma Physics and Magnetohydrodynamics and Their Applications to Controlled Thermonuclear Reactions (Univ. of Maryland Engineering & Physical Science Library).
211. Rohrlich, F. (1960), The Classical Electron, in Lectures on Theoretical Physics (Interscience).
212. Rosenfeld, L. (1951), Theory of Electrons (North-Holland).
213. Sagdeev, R. Z., and Z. A. Galeev (1969), Nonlinear Plasma Theory (Benjamin).
214. Schmidt, G. (1966), Physics of High Temperature Plasma (Academic Press).
215. Shapiro, A. H. (1953), Compressible Fluid Flow, vol. I, II (Ronald Press).
216. Simon, A. (1959), An Introduction to Thermonuclear Research (Pergamon Press).
217. Spitzer, L. (1956), Physics of Fully Ionized Gas (Interscience).
218. Steele, M. C., and B. Vural (1969), Wave Interactions in Solid State Plasmas (McGraw-Hill).
219. Sutton, G. W., and A. Sherman (1965), Engineering Magnetohydrodynamics (McGraw-Hill).
220. Tatarski, V. I. (1961), Wave Propagation in Turbulent Media (McGraw-Hill).
221. Trivelpiece, A. W. (1967), Slow Wave Propagation in Plasma Waveguides (San Francisco Press).
222. Vandenplas, P. E. (1968), Electron Waves and Resonances in Bounded Plasmas (Wiley-Interscience).
223. Vedenov, A. A. (1966), Theory of Turbulent Plasma, NASA-TT-F-436.
224. Wait, J. R., and E. A. Brackett (1967), Bibliography on Waves in Plasmas, Tech. Rep. IER 39-ITSA 39, ESSA, Boulder, Colorado.
225. Zel'dovich, Ya. B., and Yu, P. Raizer (1967), Physics of Shock Waves and High Temperature Hydrodynamic Phenomena, vol. I, II (Academic Press).

G. PLASMA PHYSICS - EDITED VOLUMES

226. Allis, W. P. (1967), Electrons, Ions and Waves (M.I.T. Press).
227. Bershader, D. (1959), Magnetodynamics of Conducting Fluids (Stanford Univ. Press).
228. Buneman, O., and W. B. Pardo (1968), Relativistic Plasmas (Benjamin).
229. Chang, C. C., and S. S. Huang (1966), Plasma Space Science Symposium (Gordon and Breach).
230. Clauser, F. H. (1960), Symposium on Plasma Dynamics (Addison-Wesley).
231. Conference (1966), Plasma Physics and Controlled Nuclear Fusion Research, 2nd., Culham, England-1965 (International Atomic Energy Agency, Vienna, Austria).
232. DeWitt, C., and J. F. Detoeuf (1960), La Theorie des Gaz Neutres et Ionises (Herman).
233. Drummond, J. E. (1961), Plasma Physics (McGraw-Hill
234. Felsen, L. B. (1967), Special Issue on Electromagnetic Wave Propagation in Anisotropic Media, Radio Science, 2, No. 8, 757-944.
235. Folkestad, K. (1968), Ionospheric Radio Communications, Proc. NATO Institute, Finse, Norway (Plenum Press).
236. Kadomtsev, B. B., M. N. Rosenbluth, and W. B. Thompson (1965), Seminar on Plasma Physics (International Atomic Energy Agency, Vienna, Austria).
237. Kash, S. W. (1959), Plasma Acceleration (Stanford University Press).
238. Kelleher, D. (1967), Eighth International Conferenc on Phenomena in Ionized Gases (Internation Atomic Energy Agency, Vienna, Austria).
239. Kunkel, W. B. (1966), Plasma Physics in Theory and Application (McGraw-Hill).
240. Landshoff, R. K. M. (1957), Magnetohydrodynamics (Stanford Univ. Press).
241. Landshoff, R. K. M. (1958), The Plasma in a Magneti Field (Stanford Univ. Press).
242. Leontovich, M. A. (1969), Reviews of Plasma Physics vol. 1-5 (Plenum Press).
243. Lochte-Holtgreven, W. (1968), Plasma Diagnostics (Wiley).
244. Longmire, C., et al. (1959), Progress in Nuclear Energy: Plasma Physics and Thermonuclear Research (Pergamon Press).
245. Maecker, H. (1962), International Conference on Ionization Phenomena in Gases, 5th, Munich (North-Holland).

246. Menzel, D. A. (1962), Selected Papers on Physical Processes in Ionized Plasmas (Dover).
247. Mitchner, M. (1961), Waves in Plasmas (Stanford Univ. Press).
248. Pai, S. I., et al. (1966), Symposium on the Dynamics of Fluids and Plasmas (Academic Press).
249. Pardo, W. B., and H. S. Robertson (1966), Plasma Instabilities and Anomalous Transport (Univ. Miami Press).
250. Perovic, B., and D. Tosic (1966), International Conference on Ionization Phenomena in Gases, 7th, Beograd (Gradevinska Knjiga Pub. House, Belgrad, Yugoslavia).
251. Rosenbluth, M. N. (1964), Advanced Plasma Theory (Academic Press).
252. Sagdeev, R. Z. (1965), Collective Processes and Shock Waves in Rarefield Plasma, Institute of Nuclear Physics, Siberian Branch, Academy of Sciences, USSR.
253. Simon, A., and W. B. Thompson (1968), Advances in Plasma Physics, vol. I (Interscience).
254. Sinelnikov, K. D. (1966), Plasma Physics and Controlled Thermonuclear Fusion, NASA-TT-F-433.
255. Sinelnikov, K. D. (1967), High Frequency Properties of Plasma, NASA-TT-F-449.
256. Skobel'tsyn, D. V. (1968), Plasma Physics (Plenum Press).
257. Wait, J. R. (1965), Special Issues on Waves in Plasmas, Radio Science J. Research NBS, 69D, Nos. 2-6.
258. Wait, J. R. (1969), Antennas in Plasma, Chapter 25 in Collin, R. E. and F. J. Zucker, Antenna Theory, Part 2, 515-556.
259. Young, F. J. (1968), Special Issue on MHD Power Generation, IEEE Proc., 56, 1407-1583.

ARTICLES

H. PROPAGATION IN MOVING MEDIA AND MOVING BOUNDARIES

260. Bailey, V. A. (1948), Plane waves in an ionized gas with static electric and magnetic fields present, Aust. J. Sci. Res., A-1, 351-359.
261. Bailey, V. A. (1950), On the relativistic electromagneto-ionic theory of wave propagation, Phys. Rev. Lett., 77, 418-419.
262. Bailey, V. A. (1951), The relativistic theory of electro-magneto-ionic waves, Phys. Rev., 83, 439-454.

263. Barsukov, K. A. (1959), On the Doppler effect in an anisotropic and gyrotropic medium, Sov. Phys. JETP, 9, 1052-1056.
264. Barsukov, K. A. (1968), Certain features of the Doppler effect in anisotropic media, Sov. Phys. Tech. Phys., 7, 112-116.
265. Barsukov, K. A., and A. A. Kolomenskii (1960), Doppler effect in an electron plasma in a magnetic field, Sov. Phys. Tech. Phys., 4, 868-870.
266. Berger, H. (1968), Reflection and transmission of electromagnetic power at moving interfaces, J. Appl. Phys., 39, 3512-3513.
267. Berger, H., and J. W. E. Griemsmann (1967), Poynting's theorem for moving media, IEEE Trans., AP-15, 490.
268. Berger, H., and J. W. E. Griemsmann (1967), Moving media without electromagnetic drag, IEEE Trans., AP-15, 585.
269. Berger, H., and J. W. E. Griemsmann (1967), Comments on guided waves in a simple moving medium, IEEE Proc., 55, 1214-1215.
270. Berger, H., and J. W. E. Griemsmann (1968), Guided waves in moving dispersive media, part I: Nonrelativistic velocities; part II: Relativistic velocities, IEEE Trans., MTT-16, 11-20.
271. Berger, H., and J. W. E. Griemsmann (1968), Transient electromagnetic guided wave propagation in moving media, IEEE Trans., MTT-16, 842-849.
272. Berger, H., and J. W. E. Griemsmann (1968), The relativistic Doppler equations for attenuated waves and the drag effect, J. Appl. Phys., 39, 3569-3573.
273. Bernstein, I. B., and S. K. Trehan (1960), Plasma Oscillations (I), Nuclear Fusion, 1, 3-41.
274. Bers, A., and P. Penfield (1962), Conservation principles of plasmas and relativistic electron beams, IRE Trans. Electron Devices, ED-9, 12-26.
275. Birkemeier, W. P., H. S. Merrill, D. H. Sargeant, D. W. Thomson, C. M. Beamer, and G. T. Bergemann (1968), Observation of wind-produced Doppler shifts in tropospheric scatter propagation, Radio Science, 3, 309-317.
276. Boffi, L. V. (1957), Electrodynamics of Moving Media Sc.D. Thesis, M.I.T., Cambridge, Mass.
277. Brandewie, R. (1963), The Interaction of Electromagnetic Waves with a Shock Heated Plasma, Ph.D. Thesis, Carnegie Inst. of Technology, Pittsburgh, Pa.

278. Brown, P. M., and C. T. Tai (1964), A study of electrodynamics of moving media, IEEE Proc., 52, 1362-1363.
279. Carstoin, J. (1967), Note on electric and magnetic polarization in moving media, Proc. Nat. Acad. Sci., 57, 1531-1541.
280. Censor, D. (1968), Relativistic Doppler broadening diagnostics, IEEE Trans., NS-15, 27-30.
281. Censor, D. (1968), First-order propagation in moving media, IEEE Trans., MTT-16, 565-566.
282. Censor, D. (1969), Propagation and scattering in radially flowing media, IEEE Trans., MTT-17, 374-378.
283. Censor, D. (1969), Scattering of a plane wave at a plane interface separating two moving media, Radio Science, 4, 1079-1088.
284. Chan, K. L., and O. G. Villard (1962), Observation of large scale travelling ionospheric disturbances by h. f. instantaneous frequency measurements, J. Geophys. Res., 67, 937-988.
285. Chan, K. L., and O. G. Villard (1963), Sudden frequency deviations induced by solar flares, J. Geophys. Res., 68, 3197-3224.
286. Chang, C. T. (1961), Shock wave phenomena in coaxial plasma guns, Phys. Fluids, 4, 1085.
287. Chen, C. P., and W. F. Hughes (1967), Electromagnetic wave interactions with moving ionizing shock, J. Geoph. Res., 72, 6021-6037.
288. Chen, H. C., and D. K. Cheng (1966), Constitutive relations for a moving anisotropic medium, IEEE Proc., 54, 62-63.
289. Chen, H. C., and D. K. Cheng (1967), A useful matrix inversion formula and its applications, IEEE Proc., 55, 705-706.
290. Cheng, D. K., and H. C. Chen (1967), On waves in an anisotropic plasma imbedded in a moving dielectric medium, IEEE Proc., 55, 1631-1632, 2197.
291. Cheng, D. K., and J. A. Kong (1968), Covariant descriptions of bianisotropic media, IEEE Proc., 56, 248-251.
292. Cheng, D. K., and J. A. Kong (1968), Time-harmonic fields in source-free bianisotropic media, J. Appl. Phys., 39, 5792-5796.
293. Chu, L. J., H. A. Haus, and P. Penfield (1966), The force density in polarizable and magnetizable fluids" IEEE Proc., 54, 920-935.
294. Cohen, M. S., K. J. Harte, and J. F. Szablya (1965), The Lorentz force, , IEEE Proc., 53, 1145.

295. Collier, J. R. (1964), Fields in a Waveguide Containing a Moving Medium , Ph.D. Thesis, Ohio State Univ., Columbus, Ohio.
296. Collier, J. R. (1964), Determination of the Parameters Governing Electromagnetic Wave Propagation in a Moving Plasma , Antenna Laboratory, Ohio State University, Report No. 1565-8, Columbus, Ohio.
297. Collier, J. R., and C. T. Tai (1964), Propagation of plane waves in lossy moving media , IEEE Trans., AP-12, 375-376.
298. Collier, J. R., and C. T. Tai (1965), Guided waves in moving media, IEEE Trans., MTT-13, 441-445.
299. Collier, J. R., and C. T. Tai (1965), Plane waves in moving media, Am. J. Phys., 33, 166-167
300. Costen, R. C., and D. Adamson (1965), Three dimensional derivation of the electrodynamics jump conditions and momentum-energy laws at a moving boundary, IEEE Proc., 53, 1181-1196.
301. Costen, R. C., and D. Adamson (1966), Electrodynamic boundary conditions at a moving boundary , IEEE Proc., 54, 399-401.
302. Daly, P. (1967), Guided waves in moving media, IEEE Trans., MTT-15, 274-275.
303. Daly, P., and H. Gruenberg (1967), Energy relations for plane waves reflected from moving media, J. Appl. Phys., 38, 4486-4489.
304. Davies, K., and D. M. Baker (1966), On frequency variations of ionospherically propagated HF radio signals, Radio Sci., 1, 545-556.
305. Davies, K., J. M. Watts, and D. M. Zacharisen (1962), A study of F2 layer effects as observed with a Doppler technique, J. Geophys. Res., 67, 601-609.
306. Du, L. J. (1965), A Study of Electrodynamics Involving Moving Media, Ph.D. Thesis, Ohio State Univ., Columbus, Ohio.
307. Du, J. L., and R. T. Compton (1966), Cutoff phenomena for guided waves in moving media", IEEE Trans., MTT-14, 358-363.
308. Dunn, D. A., R. W. Wallace, and S. D. Choi (1969), Waves in a moving medium with finite conductivity, IEEE Proc., 57, 45-57.
309. Eggimann, W. H. (1965), On the electric and magnetic field equations , IEEE Proc., 53, 1642.
310. Fainberg, Ia. B., and V. S. Tkalich (1959), On the reflection of an electromagnetic wave from the plasma moving through a dielectric in a constant magnetic field, Sov. Phys. Tech Phys., 4, 438-443.

311. Fedorchenko, A. M. (1967), The theory of cyclotron waves in cold, moving plasma, Radiotekhnika i Electronica (USSR), 12, 1336-1339.
312. Fejer, J. A. (1963), Hydromagnetic reflection and refraction at a fluid velocity discontinuity, Phys. Fluids, 6, 508-512.
313. Frank, I. (1943), Doppler effect in a refractive medium, J. Phys. USSR, 7, 49-67.
314. Fujioka, H., K. Yoshida, and N. Kumagai (1967), Transmission line treatment of waveguides filled with a moving medium, IEEE Trans., MTT-15, 265-267.
315. Fujita, H., and H. Muramatsu (1968), Reflection and transmission of a plane electromagnetic wave at a fast moving boundary, IEEE Proc., 56, 1605-1606.
316. Gintsburg, M. A. (1965), Radio-wave propagation in moving cosmic plasma, Cosmic Res., 3, 251-253.
317. Goldstein, S. (1961), Concerning the Continuum Theory of the Electrodynamics and Dynamics of Moving Media, Proc. Symp. on Electromagnetics and Fluid Dynamics of Gaseous Plasma, Polytechnics Press of Polytechnic Institute Brooklyn, 65-80, Brooklyn, N.Y.
318. Gross, R. A. (1965), Strong ionization shock waves , Rev. Mod. Phys., 37, 724-743.
319. Gruenberg, H., and P. Daly (1967), Waveguides containing moving dispersive media , IEEE Trans., MTT-15, 636-642.
320. Guery, F. (1924), Les lois de la reflexion sur des miroirs en mouvement, General Electric Review, 16, 371-376.
321. Harris, E. G. (1957), Relativistic magnetohydrodynamics, Phys. Rev., 108, 1357-1360.
322. Hines, C. O. (1959), Motions in the ionosphere," IRE Proc., 47, 176-186.
323. Hughes, W. F., and F. J. Young (1964), The interaction of transverse electromagnetic plane waves and a moving ionizing shock wave in the presence of a magnetic field, J. Fluid Mech., 19, 11-20.
324. Jacobs, J. A., and T. Watanabe (1966), Doppler frequency changes in radio waves propagating through a moving ionosphere , Radio Science, 1, 257-264.
325. Kahalas, S. L., and D. A. McNeill (1964), Coupling of magnetohydrodynamic to electromagnetic acoustic waves at a plasma-neutral gas interface, Phys. Fluids, 7, 1321-1327.

326. Kalafus, R. M. (1967), Electrodynamics of moving conducting media, Ph.D. Thesis, Univ. of Michigan, Ann Arbor, Michigan.

327. Kalafus. R. M. (1968), Electromagnetism in moving conducting media, NASA Tech. Rep. CR-1025.

328. Kanellakos, O. P. (1963), Origin and location of ionospheric perturbations affecting the instantaneous frequency and azimuthal angle of arrival of HF waves, Radio Astron. and Satel. Studies of the Atm., J. Aarons, Ed. (North-Holland) 525-559.

329. Kerzar, B., and P. Weissglas (1965), Determination of electron drift velocity in a gaseous discharge from Doppler shifts of plasma waves, Electronics Letters, 1, 43-44.

330. Kimura, I., and R. Nishina (1967), Doppler shift in an inhomogeneous anisotropic medium, Rep. Ionosphere Space Res. Japan, 21, 187-192.

331. Knox, F. B. (1962), Reflection of an electromagnetic wave from a moving boundary between two independently moving ionized media, J. Atmos. Terr. Phys., 24, 1003-1010.

332. Kong, J. A., and D. K. Cheng (1968), Wave reflections from a conducting surface with a moving uniaxial sheath, IEEE Trans., AP-16, 577-583.

333. Kong, J. A., and D. K. Cheng (1968), On guided waves in moving anisotropic media, IEEE Trans. MTT-16, 99-103.

334. Kong, J. A., and D. K. Cheng (1968), Wave behavior at an interface of a semi-infinite moving anisotropic medium, J. Appl. Phys., 39, 2282-2286.

335. Kong, J. A., and D. K. Cheng (1969), Scattering from a conducting cylinder coated with a moving uniaxial medium, Canad. J. Phys. 47, 353-360.

336. Kong, J. A., and D. K. Cheng (1969), Reflection and refraction of electromagnetic waves by a moving uniaxially anisotropic slab, J. Appl. Phys. 40, 2206-2212.

337. Kooy, C. (1969), Free current vibrations in wire antennas embedded in a moving medium, IEEE Trans., AP-17, 820-821.

338. Kouyoumjian, R. G. (1968), On the electromagnetic power conversion in problems involving moving boundaries, IEEE Proc., 56, 2185-2186.

339. Koyama, M., and K. Itakura (1967), Unified theory of electrodynamics of moving media, J. Inst. Elect. Commun. Engrs. Japan, 50, 1940-1947.

340. Kritikos, H. N. (1967), The eikonal equation in a moving medium, IEEE Proc., 55, 442-443.
341. Kritikos, H. N., K. S. H. Lee, and C. H. Papas (1967), Electromagnetic reflectivity of non-uniform jet streams, Radio Science, 2, 991-995.
342. Kurilko, V. I. (1961), Kinetic theory of reflection of electromagnetic waves from a moving plasma, Sov. Phys. Tech. Phys., 6, 50-54.
343. Kurilko, V. I., and V. I. Miroshnichenko (1963), Reflection of electromagnetic waves from a moving plasma, Sov. Phys. Tech. Phys., 7, 588-592.
344. Lam, J. (1967), Scattering by an expanding conducting sphere, Antenna Lab., California Inst. Tech., Tech. Rep. 40, Pasadena, California.
345. Lee, K. S. H., and C. H. Papas (1963), Doppler effects in inhomogeneous anisotropic ionized gases, J. Math. Phys., 4, 189-199.
346. Lee, S. W., and Y. T. Lo (1967), Reflection and transmission of electromagnetic waves by a moving uniaxially anisotropic medium, J. Appl. Phys., 38, 870-875.
347. Lee, S. W., and R. Mittra (1967), Scattering of electromagnetic waves by a moving cylinder in free space, Canad. J. Phys., 45, 2999-3007.
348. Lerche, I. (1966), Transverse waves in a relativistic plasma, Phys. Fluids, 9, 1073-1080.
349. Lichtenberg, A. J. (1962), Dispersion relations for quasi static waves on moving finite cross-section plasma, Electronic Research Lab., Univ. of California, Scientific Report No. 22, Berkeley, California.
350. Liperovskii, V. A. (1961), Anisotropic propagation of longitudinal electro-acoustical waves in drifting plasma, Sov. Phys. JETP, 12, 951-953.
351. Longo, S. E., J. L. Mills, and Y. J. Seto (1967), The reflection coefficient for electromagnetic waves incident upon a homogeneous moving plasma in a shock tube, Southwest IEEE Conference Record, April 19-21, 1967, Dallas, Texas, paper No. 2-5-1.
352. Lusignan, B. B. (1963), Detection of solar particle streams by high frequency radio waves, J. Geophys. Res., 68, 5617-5632.
353. Manz, B. (1967), Doppler shift in a plasma, J. Opt. Soc. Amer., 57, 1543-1550.
354. McKenzie, J. F. (1967), Electromagnetic waves in uniformly moving media, Proc. Phys. Soc., 91, pt. 3, 352-356.

355. Mimmo, H. P. (1937), The physics of the ionosphere, Rev. Mod. Phys., 9, 1-43.

356. Minkowski, H. (1908), Die grundgleichungen fur die elektromagnetischen vorgange in bewegten korpern, Gottingen Nachzichten, 53-116.

357. Minkowski, H. (1910), Eine ableitung der grundgleichungen fur die elektromagnetischen vorgange in bewegten korpern vom standpunkte der elektronentheorie, Math. Ann., 68, 526-551.

358. Munro, G. H. (1958), Traveling ionospheric disturbances in the F-region, Australian J. Phys., 11, 91-112.

359. Murphy, J. H., and W. F. Hughes (1966), Observations of magnetoacoustic waves in mercury, IEEE Proc., 54, 55-56.

360. Nathan, A., and D. Censor (1967), Extended ∇ relations with reference to EM waves in moving simple media, Faculty of Electrical Engineering, Israel Institute of Technology, Pub. No. EE-72, Haifa, Israel.

361. Nathan, A., and D. Censor (1968), Extended ∇ relations with reference to EM waves in moving simple media, IEEE Trans., MTT-16, 883-884.

362. Oster, L. (1960), Linearized theory of plasma oscillations, Rev. Mod. Phys., 32, 141-169.

363. Otsuka, M. (1966), Doppler-effect-like phenomena, J. Phys. Soc. Japan, 21, 1398-1410.

364. Pavlovskii, R. A. (1966), Some approximations in the electrodynamics of moving media, Magn. Gidrodinamika (USSR), 3, 146-148.

365. Peer, R. F. (1965), An Application of the Lorentz Transformation to Cylindrical Waveguides, M.S. Thesis, Washington Univ., St. Louis, Missouri.

366. Penfield, P., and H. A. Haus (1965), Electrodynamics of moving media, IEEE Proc., 53, 422.

367. Penfield, P., and H. A. Haus (1966), Electromagnetic force density, IEEE Proc., 54, 328.

368. Penfield, P., and H. A. Haus (1966), Hamilton's principle for electromagnetic fluids, Phys. Fluids, 9, 1195-1204.

369. Penfield, P., and J. F. Szablya (1965), The Lorentz force, IEEE Proc., 53, 1144-1145.

370. Penfield, P., and C. T. Tai (1964), Electromagnetism of moving media, IEEE Proc., 52, 1361-1362.

371. Podolsky, B., and H. Denman (1955), A derivation of generalized macroscopic electrodynamics equations, I. Non-relativistic, J. Math. Phys., 34, 198-207.

372. Pyati, V. P. (1967), Reflection and refraction of electromagnetic waves by a moving dielectic medium, J. Appl. Phys., 38, 652-655.

373. Ramasastri, J., and G. Y. Chen (1967), Wave interaction with moving boundaries, Electronics Letters, 3, 479-481.

374. Restrick, R. C. (1967), Scattering by moving bodies, Ph.D. Thesis, Univ. of Michigan, Ann Arbor, Michigan.

375. Restrick, R. C. (1968), Electromagnetic scattering by a moving conducting sphere, Radio Science, 3, 1144-1154.

376. Risco, M. (1947), Le principe de Huyghens dans l'optique des corps en mouvement, J. Phys. Radium, 8, 282-288.

377. Risco, M. (1949), Ondes planes et ondes spheriques dans les problemes optiques avec mouvement relatif. Cas d'un miroir mobile illumine par un faisceau convergent, J. Phys. Radium, 10, 128-131.

378. Risco, M. (1950), Etude theorique de la formation d'images mobiles ponctuelles. Role cinematique des conditions d'Abbe et de Herschell, J. Phys. Radium, 11, 159-163.

379. Rowlands, J., V. L. Sizonenko, and K. N. Stepanov (1966), Contribution to the theory of decay of electromagnetic waves in a magnetoactive plasma, Soviet Phys. JETP, 23, 661-667.

380. Rydbeck, O. E. H. (1964), Research in the field of propagation of high power waves in ionized media, Research Laboratory of Electronics, Chalmers University of Technology, Research (Survey) Report No. 48, Gothenburg, Sweden.

381. Sakuntala, M., B. E. Clotfelter, W. B. Edwards, and R. G. Fowler (1959), Electromotive force in a highly ionized plasma moving across a magnetic field, J. Appl. Phys., 30, 1669-1671.

382. Scarf, F. L. (1961), Wave propagation in a moving plasma, Am. J. Phys., 29, 101-107.

383. Sharma, S. R. (1968), Wave propagation in a plasma drifting in a magnetic field, Physica, 38, 625-632.

384. Shiozawa, T. (1966), Guided waves in a simple moving medium, IEEE Proc., 54, 1984-1985.

385. Shiozawa, T., and K. Hazama (1968), General solution to the problem of reflection and transmission by a moving dielectric medium, Radio Science, 3, 569-576.

386. Shiozawa, T., K. Hazama, and N. Kumagai (1967), Reflection and transmission of electromagnetic waves by a dielectric half-space moving perpendicular to the plane of incidence, J. Appl. Phys., 38, 4459-4461.

387. Shiozawa, T., and N. Kumagai (1967), Total reflection at the interface between relatively moving media, IEEE Proc., 55, 1243-1244.

388. Silin, V. P. (1961), Electromagnetic properties of a relativistic plasma, Sov. Phys. JETP, 13, 430.

389. Silin, V. P., and Fetisov, E. P. (1962), Electromagnetic properties of a relativistic plasma, Sov. Phys. JETP, 14, 115.

390. Smy, P. R., and A. Offenberger (1963), Experiments with a moving magnetized plasma, Canad. J. Phys., 41, 469.

391. Stolyarov, S. N. (1963), Reflection and refraction of electromagnetic waves at a moving boundary, Sov. Phys. Tech. Phys., 8, 418-422.

392. Stolyarov, S. N. (1967), Reflection and passage of electromagnetic wave incident on moving dielectric plate, Izv. VUZ Radiofiz. (USSR), 10, 284-288.

393. Szablya, J. F. (1965), On the Lorentz force, IEEE Proc., 53, 418.

394. Tai, C. T. (1964), A study of electrodynamics of moving media, IEEE Proc., 52, 685-689.

395. Tai, C. T. (1964), On the electrodynamics in the presence of moving matter, IEEE Proc., 52, 307-308.

396. Tai, C. T. (1965), Electrodynamics of moving anisotropic media: The first order theory, Radio Sci. J. Res., NBS 69D, 401-405.

397. Tai, C. T. (1965), Reflection and refraction of a plane electromagnetic wave at the boundary of a semi-infinite moving medium, Radiation Laboratory, University of Michigan, Report No. RL-302, Ann Arbor, Michigan.

398. Tai, C. T. (1965), Definite and indefinite forms of Maxwell's equations for moving media, Radiation Laboratory, University of Michigan, Report No. RL-310, Ann Arbor, Michigan

399. Tai, C. T. (1967), Present views on electrodynamics of moving media, Radio Science, 2, 245-248.

400. Tai, C. T., and R. J. Plugge (1964), Nomographs for relativistic velocity transformation, IEEE Trans. Education, E-7, 145-148.

401. Tai, C. T., and J. F. Szablya (1965), Comments on 'On the Lorentz Force, IEEE Proc., 53, 1145

402. Tan, T. S. (1968), Electromagnetic Wave Propagation in Two Relatively Moving Media, M.S. Thesis, Polytechnic Inst. of Brooklyn, Brooklyn, N.Y.
403. Tang, C. L., and J. Meixner (1961), Relativistic theory of the propagation of plane electromagnetic waves , Phys. Fluids, 4, 148-154.
404. Tsai, C. S. (1968), Scattering of electromagnetic waves from a moving layer in a transmission line, J. Appl. Phys., 39, 133-136.
405. Tsai, C. S., and B. A. Auld (1967), Wave interactions with moving boundaries, J. Appl. Phys., 38, 2106-2115.
406. Tsandoulas, G. N. (1968), On the Doppler effect in plane wave scattering by moving conducting bodies, IEEE Proc., 56, 1749-1750.
407. Tsandoulas, G. N. (1968), Electromagnetic diffraction by a moving wedge, Radio Science, 3, 887-893.
408. Valenzuela, G. R. (1968), Scattering of electromagnetic waves from a slightly rough surface moving with uniform velocity, Radio Science, 3, 1154-1157.
409. Vezzetti, D. J., and J. B. Keller (1967), Refractive index, attenuation, dielectric constant and permeability for waves in polarizable medium, J. Math. Phys., 8, 1861-1870.
410. Wei, C. C. (1959), Relativistic hydrodynamics for a charged non-viscous fluid, Phys. Rev., 113, 1414.
411. Yeh, C. (1965), Reflection and transmission of electromagnetic waves by a moving dielectric medium, J. Appl. Phys., 36, 3513-3517.
412. Yeh, C. (1966), A proposed method of shifting the frequency of light waves, Appl. Phys. Lett., 9, 184-185.
413. Yeh, C. (1966), Reflection and transmission of electromagnetic waves by a moving plasma medium, J. Appl. Phys., 37, 3079-3082.
414. Yeh, C. (1967), Reflection from a dielectric-coated moving mirror, J. Opt. Soc. Am., 57, 657-661.
415. Yeh, C. (1967), Reflection and transmission of electromagnetic waves by a moving plasma medium, II. Parallel polarization, J. Appl. Phys., 38, 2871-2873.
416. Yeh, C. (1967), Brewster angle for a dielectric medium moving at relativistic speed, J. Appl. Phys., 38, 5194-5200.

417. Yeh, C. (1968), Propagation along moving dielectric waveguides, J. Opt. Soc. Am., 58, 767-770.

418. Yeh, C. (1968), Reflection and transmission of electromagnetic waves by a moving dielectric slab. II. Parallel polarization, Phys. Rev., 167, 875-877.

419. Yeh, C. (1969), Scattering obliquely incident microwave by a moving plasma column, J. Appl. Phys., 40, 5066-5075.

420. Yeh, C. (1970), Diffraction of waves by a conducting cylinder coated with a moving plasma sheath, J. Math. Phys., 11, 99-104.

421. Yeh, C., and K. F. Casey (1966), Reflection and transmission of electromagnetic waves by a moving dielectric slab, Phys. Rev., 144, 665-669.

422. Young, F. J., R. C. Costen, and D. Adamson (1966), Electrodynamic boundary conditions at a moving boundary, IEEE Proc., 54, 399-401.

I. RADIATION IN MOVING MEDIA AND CERENKOV RADIATION

423. Abele, M. (1962), Radiation in a plasma from a uniformly moving distribution of electric charge, Proc. Symp. on Electromagnetics and Fluid Dynamics of Gaseous Plasma, Polytechnic Institute of Brooklyn, 153-17

424. Bazhanova, A. E. (1965), Complex Doppler effect in plasma, Soviet Radiophysics, 8, 795-803.

425. Bergeson, J. E. (1968), Radiation from an oscillating magnetic dipole in streaming plasma, Radio Science, 3, 191-200.

426. Besieris, I. M. (1967), The time-dependent Green's function for electromagnetic radiation in a moving conducting medium: nonrelativistic approximation, J. Math. Phys., 8, 409-416.

427. Besieris, I. M., and R. T. Compton (1967), Time dependent Green's function for electromagnetic waves in moving conducting media, J. Math. Phys., 8, 2445-2451.

428. Bolotovskii, B. M., and A. K. Burtsev (1965), Radiation of a charge moving above a diffraction grating, Optics and Spectrosc., 19, 263-265.

429. Bolotovskii, B. M., and A. A. Rukhadze (1960), Field of a charged particle in a moving medium, Sov. Phys. JETP, 10, 958-961.

430. Casey, K. F., and C. Yeh (1969), Radiation from an aperture in a conducting cylinder coated with a moving plasma sheath, IEEE Trans., AP-17, 757-762.

431. Censor, D. (1967), Scattering in velocity dependent systems, D. Sc. Thesis, Technion-Israel Institute of Technology, Haifa, Israel (In Hebrew).

432. Censor, D. (1968), Scattering by moving objects and scattering in moving media, Proc. 6th Nat. Conv. of Elec. and Electronics Engrs. in Israel (Tel-Aviv, October 20-22, 1968), 466-480.

433. Censor, D. (1969), Scattering of electromagnetic waves by a cylinder moving along its axis, IEEE Trans., MTT-17, 154-158.

434. Censor, D., and A. Nathan (1969), Scattering by a rotating cylinder, Faculty of Electrical Engineering, Israel Institute of Technology, Pub. No. EE-116, Haifa, Israel.

435. Cheng, D. K., and F. I. Tseng (1968), Theorems concerning optimization and synthesis of antenna arrays in a moving medium, IEEE Proc., 56, 1622-1623.

436. Cohen, M. (1961), Radiation in a plasma, I. Cerenkov effects, Phys. Rev., 123, 711-721.

437. Cohen, M. H. (1962), Radiation in a plasma, II. equivalent sources, Phys. Rev., 126, 389-397.

438. Cohen, M. H. (1962), Radiation in a plasma, III. metal boundaries, Phys. Rev., 126, 398-403.

439. Coleman, P. D. (1964), Cerenkov radiation and allied phenomena, Proc. Symp. on Quasi Optics, Polytechnic Institute of Brooklyn, 199-216.

440. Compton, R. T. (1966), The time dependent Green's function for electromagnetic waves in moving simple media, J. Math. Phys., 7. 2145-2152.

441. Compton, R. T. (1968), One and two-dimensional Green's functions for electromagnetic waves in moving simple media, J. Math. Phys., 9, 1865-1872.

442. Compton, R. T., and C. T. Tai (1964), Poynting's theorem for radiating systems in moving media, IEEE Trans., AP-12, 238-239.

443. Compton, R. T., and C. T. Tai (1964), The dyadic Green's function for an infinite moving medium, Antenna Lab., Ohio State University, Tech. Rep. 1691-3, Columbus, Ohio.

444. Compton, R. T., and C. T. Tai (1965), Radiation from harmonic sources in a uniformly moving medium, IEEE Trans., AP-13, 574-577.

445. Daly, P., K. S. H. Lee, and C. H. Papas (1965), Radiation resistance of an oscillating dipole in a moving medium, IEEE Trans., AP-13, 583-587.

446. Dolginov, A. Z., and I. Toplygin (1967), Multiple scattering of charges in magnetic field with moving random inhomogeneities, Soviet Phys. JETP, 24, 1195-1202.

447. Eidman, V. Ia (1958), The radiation from an electron moving in a magneto plasma, Sov. Phys. JETP, 7, 91-95.

448. Eidman, V. Ia (1962), Radiation of plasma waves by a charge moving in a magnetoactive plasma, Sov. Phys. JETP, 14, 1401.

449. Felsen, L. B., and A. Hessel (1961), A network approach to the analysis of Cerenkov radiation problems, Il Nuovo Cimento, B, 19, 1065-1071.

450. Francia, di G. T. (1960), On the theory of some Cerenkovian effects, Il Nuovo Cimento, B, 16, 61-77.

451. Frank, I. M., and I. Tamm (1937), Coherent visible radiation from fast electrons passing through matter, Dokl. Akad. Nauk. SSSR, 3, 109-114.

452. Fujioka, H., and N. Kumagai (1966), Radiation characteristics of antenna arrays in a moving medium, Electronics and Commun. Japan, 49, 100-107.

453. Fujioka, H., and N. Kumagai (1967), Electromagnetic radiation in a moving dissipative medium, J. Inst. Elec. Commun. Engrs. Japan, 50, 1906-1910.

454. Fujioka, H., and N. Kumagai (1967), Electromagnetic radiation in a moving lossy medium, Radio Science, 2, 1449-1458.

455. Fujioka, H., T. Shiozawa, and N. Kumagai (1966), Electromagnetic radiation from an electric dipole moving with relativistic velocity, Electronics and Commun. Japan, 49, 100-107.

456. Fung, P. C. W. (1966), Excitation of backward Doppler-shifted cyclotron radiation in a magneto-active plasma by an electron stream, Planet. Space Sci., 14, 335-346.

457. Ginzburg, V. L., and I. M. Frank (1947), The emission from an electron of an atom moving along the axis of a channel in a dense medium, Dokl. Akad. Nauk, USSR, 56, 699-702.

458. Gluckman, A. G. (1967), Historical note on the symmetry of electrodynamic processes of electrified bodies in motion, IEEE Proc., 55, 123-124.
459. Handelsman, R., and R. M. Lewis (1966), Asymptotic theory of Cerenkov radiation, J. Math. Phys., 7, 1982-1999.
460. Hazama, K., T. Shiozawa, and I. Kawano (1969), Effect of a moving dielectric half-space on the radiation from a line source, Radio Science, 4, 483-488.
461. Holmes, J. F., and A. Ishimaru (1969), Relativistic communications effects associated with moving space antennas, IEEE Trans., AP-17, 484-488.
462. Jauch, J. M., and K. M. Watson (1948), Phenomenological quantum-electrodynamics, Phys. Rev., 74, 950-957.
463. Jauch, J. M., and K. M. Watson (1948), Phenomenological quantum-electrodynamics, Phys. Rev., 74, 1485-1493.
464. Johnson, P. S. (1962), Cerenkov radiation spectra for a cold magneto-active plasma, Phys. of Fluids, 5, 118-120.
465. Kahalas, S. L. (1963), Coupling of magnetohydrodynamic to electromagnetic waves at a plasma discontinuity: II--the nonpropagating field, Phys. Fluids, 6, 438.
466. Kalafus, R. M. (1969), Radiation from sources in a moving, conducting medium, Radio Science, 4, 977-982.
467. Kenyon, R. (1962), Cerenkov radiation in a plasma, University of Illinois, Tech. Rep. No. 4, Contract No. AF33 (616)-7043, Urbana, Ill.
468. Kolomenskii, A. A. (1956), Radiation from a plasma electron in uniform motion in a magnetic field, Sov. Phys. - Doklady, 1, 133-136.
469. Kolpakov, O. A., and V. I. Kotov (1966), The radiation for a magnetic dipole moving through a cylindrical resonator and a structural waveguide, Soviet Phys. Tech. Phys., 10, 1649-1651.
470. Kolsrud, M., and E. Leer (1967), Radiation from moving dipoles, Phys. Norveg., 2, 181-188.
471. Lampert, M. A. (1956), Incidence of an electromagnetic wave on a Cerenkov electron gas, Phys. Rev., 102, 299-340.
472. Lashinsky, H. (1961), Advances in Electronics and Electron Physics (Academic Press), 14, 265-297.

473. Lee, K. S. H. (1968), Radiation from an oscillating source moving through a dispersive medium with particular reference to the complex Doppler effect, Radio Science, 3, 1098-1104.

474. Lee, K. S. H., and C. H. Papas (1964), Electromagnetic radiation in the presence of moving simple media, J. Math. Phys., 5, 1668-1672

475. Lee, K. S. H., and C. H. Papas (1965), Antenna radiation in a moving dispersive medium, IEEE Trans., AP-13, 799-804.

476. Lee, S. W., and Y. T. Lo (1966), Radiation in a moving anisotropic medium, Radio Science, 1, 313-324.

477. Lewis, R. M., and W. Pressman (1966), Radiation and propagation of electromagnetic waves from moving sources, Radio Science, 1, 1029-104

478. Lewis, R. M., and W. Pressman (1967), The progressing wave formalism for electromagnetic waves from moving sources, J. Diff. Eq., 3, 360-390.

479. Linhart, J. G. (1955), Cerenkov radiation of electrons moving parallel to a dielectric boundary, J. Appl. Phys., 26, 527-534.

480. Majundar, S. K. (1960), Electrodynamics of a charged particle moving through a plasma without magnetic field, Proc. Phys. Soc. (London), 76, 657-669.

481. Majundar, S. K. (1961), Radiation by charged particles passing through an electron plasma in an external magnetic field, Proc. Phys. Soc. (London), 77, 1109-1120.

482. McKenzie, J. F. (1963), Cerenkov radiation in a magneto-ionic medium, Phil. Trans. Roy. Soc. (London), A, 255, 585-606.

483. McKenzie, J. F. (1967), Dipole radiation in moving media, Proc. Phys. Soc., 91, Pt. 3, 537-551.

484. McKenzie, J. F. (1967), Effect of the motion of a strongly magnetized plasma on the emission of radiation of a finite dipole, J. Appl. Phys., 38, 5249-5255.

485. Meyers, N. H. (1958), A Poynting theorem for moving bodies and the relativistic mechanics of extended objects, J. Franklin Institute, 266, 439-464.

486. Morozov, A. I. (1958), The use of Leontovich's boundary conditions in the theory of Cerenkov radiation, Sov. Phys. JETP, 6, 717-718.

487. Mott, H., G. B. Hoadley, and W. R. Davis (1967), Radiation from a moving electric dipole, IEEE Proc., 55, 92-93.

488. Motz, H. (1951), Applications of the radiation from fast electron beams, J. Appl. Phys., 22, 527-535.

489. Nag, B. D., and A. M. Sayied (1956), Electrodynamics of moving media and the theory of Cerenkov effect, Proc. Roy. Soc. (London), A, 235, 544-551.

490. Newburgh, R. G. (1968), Radiation and the classical electron, Am. J. Phys., 36, 399-405.

491. Ott, R. H., and G. Hufford (1968), Scattering by an arbitrarily shaped conductor in uniform motion relative to the source of an incident spherical wave, Radio Science, 3, 857-861.

492. Pafomov, V. E. (1956), Cerenkov radiation in anisotropic ferrites, Sov. Phys. JETP, 3, 597-600.

493. Pafomov, V. E. (1957), Peculiarities of Cerenkov radiation in anisotropic media, Sov. Phys. JETP, 5, 307-309.

494. Pafomov, V. E. (1959), Transition radiation and Cerenkov radiation, Sov. Phys. JETP, 9, 1321-1324.

495. Palocz, I., and A. A. Oliner (1965), Leaky space charge waves, I: Cerenkov radiation, IEEE Proc., 53, 24-36.

496. Pavlenko, Yu. G. (1966), Doppler frequency shift of an emitter moving in the ionosphere, Geomagn. i. Aeronomiya (USSR), 6, 790-791.

497. Pierce, J. R. (1955), Interaction of moving charges with wave circuits, J. Appl. Phys., 26, 627-638.

498. Pyati, V. (1966), Radiation from oscillating dipole above a moving half-space, Ph.D. Thesis, Univ. of Michigan, Ann Arbor, Michigan.

499. Pyati, V. (1966), Radiation due to an infinite oscillating dipole over a lossless semi-infinite moving dielectric medium, Univ. Michigan, Radiation Lab. Rept. No. 7322-2-T.

500. Pyati, V. P. (1967), Radiation due to a dipole over a moving medium, J. Appl. Phys., 38, 4372-4374.

501. Sasiela, R., and J. P. Freidberg (1967), Utilization of the refractive index surfaces to evaluate Cerenkov radiation in an infinite magnetoplasma, Radio Science, 2, 703-710.

502. Seshadri, S. R. (1967), Cerenkov radiation in a magnetoionic medium, Electronic Letters, 3, 271-274.

503. Seshadri, S. R., and H. S. Tuan (1965), Radiation from a uniformly moving charge in an anisotropic two component plasma, Radio Sci. J. Res. NBS, 69D, 767-783.

504. Seto, Y. J. (1967), The Green's dyadic for radiation in a bounded simple moving medium, IEEE Trans., MTT-15, 455-462.

505. Shiozawa, T., and N. Kumagai (1967), Radiation characteristics of a linear antenna in moving media, Electron. Commun. Japan, 50, 163-167.

506. Sitenko, A. G., and A. A. Kolomenskii (1956), Motion of a charged particle in an optically active anisotropic medium I, Sov. Phys. JETP, 3, 410-416.

507. Sollfrey, W., and H. T. Yura (1965), Cerenkov radiation from charged particles in a plasma in a magnetic field, Phys. Rev. 139, A48-A55.

508. Sugai, I. (1967), Energy-transport velocity under Lorentz transformation, IEEE Proc., 55, 727-728.

509. Sugai, I. (1968), Energy transport velocity in lossy moving fluids, IEEE Proc., 56, 1133-1134.

510. Sugihara, R. (1963), Harmonic structure of the radiation from a large moving through a magneto-plasma, J. Phys. Soc. Japan, 18, 1062-1065.

511. Sugihara, R., and Y. Itikawa (1965), Energy losses by a sequence of fast charges and by a gyrating charge in a magneto-plasma, J. Phys. Soc. Japan, 20, 591-597.

512. Tai, C. T. (1964), Two Scattering Problems Involving Moving Media, Antenna Lab., Ohio State Univ., Tech. Rep. 1691-7, Columbus, Ohio.

513. Tai, C. T. (1965), The dyadic Green's function for a moving isotropic medium, IEEE Trans., AP-13, 322-323.

514. Tai, C. T. (1965), Radiation in Moving Media, Proceeding of the Ohio State University Short Course on Recent Advances in Antenna and Scattering Theory, Columbus, Ohio.

515. Tai, C. T. (1965), Huygen's principle in a moving, isotropic, homogeneous and linear medium, Appl. Optics, 4, 1347-1349.

516. Tai, C. T. (1967), Time-dependent Green's function for a moving isotropic medium, J. Math. Phys., 8, 646-647.

517. Tret'Yakov, O. A., E. I. Chernyakov, and V. P. Shestopalov (1966), Electromagnetic wave radiation from an electron beam moving over a diffraction grating, Soviet Phys. Tech. Phys., 11, 22-25.
518. Tseng, F. I., and D. K. Cheng (1969), Pattern synthesis of circular arrays in a moving medium, IEEE Trans., AP-17, 524-526.
519. Tsintsadze, N. L., and A. D. Pataraya (1961), Cerenkov generation of hydromagnetic and magneto-acoustic waves in a rarefied anisotropic plasma, Sov. Phys. Tech. Phys., 5, 1113-1121.
520. Tsytovich, V. N. (1962), Radiative correction for the intensity of Cerenkov radiation of charged particles, Sov. Phys. Doklady, 7, 411.
521. Tsytovich, V. N. (1962), Effect of radiation on a charged particle moving through a magneto-active plasma, Sov. Phys. JETP, 16, 234.
522. Tuan, H. S., and S. R. Seshadri (1963), Radiation from a uniformily moving charge in an anisotropic plasma, IEEE Trans., MTT-11, 462-471.
523. Tuan, H. S., and S. R. Seshadri (1965), Radiation from a uniformly moving distribution of electric charge in an anisotropic, compressible plasma, IEEE Trans., AP-13, 71-78.
524. Ullah, N., and S. L. Kahalas (1963), Coupling of magnetohydrodynamic and electromagnetic waves at a plasma discontinuity, I. Radiation field, Phys. Fluids, 6, 284.
525. Watson, K. M., and J. M. Jauch (1949), Phenomenological quantum electrodynamics, III Dispersion, Phys. Rev. 75, 1249-1261.
526. Wolfe, P., and R. M. Lewis (1966), Progressing waves radiated from a moving point source in an inhomogeneous medium, J. Diff. Eq., 2, 328-350.
527. Yadavalli, S. V. (1965), Electromagnetic wake of a charge-particle pulse in a plasma, Phys. Fluids, 8, 956-967.
528. Yildiz, M., and S. N. Samaddar (1963), A note on wave phenomena due to a moving and oscillating dipole in a compressible plasma column, Canad. J. Phys., 41, 2268-2275.

J. AMPLIFICATION OF WAVES IN PLASMAS

529. Bailey, V. A. (1948), Spontaneous waves in discharge tubes and in the solar atmosphere, Nature, 161, 599-660.

530. Bailey, V. A. (1950), The growth of circularly polarized waves in the sun's atmosphere and their escape into space, Phys. Rev., 78, 428-443.
531. Bell, T., and R. A. Helliwell (1960), Travelling wave amplification in the ionosphere, Proc. Symp. on Physical Processes in the Sun-Earth Environment, D. R. B., Canada, 215-222.
532. Bohm, D., and E. P. Gross (1949), Theory of plasma oscillations, A. Origin of medium like behavior, Phys. Rev., 75, 1851-1864.
533. Bohm, D., and E. P. Gross (1949), Theory of plasma oscillations, B. Excitation and damping of oscillations, Phys. Rev., 75, 1864-1876
534. Clemmow, P. C. (1962), Wave amplification in a plasma stream in a medium of high refractive index, Proc. Phys. Soc. (London), 80, 1322-1332.
535. Clemmow, P. C. (1968), A note on the two-stream instability, J. Plasma Phys., Pt. 1, 2, 85-91.
536. Fung, P. C. W. (1966), Excitation of backward Doppler-shifted cyclotron radiation in a magneto-active plasma by an electron stream, Planet, Space Sci., 14, 335-346.
537. Getmantsev, G. G. (1960), Growth of electromagnetic waves in interpenetrating infinite moving media, Sov. Phys. JETP, 10, 600-602.
538. Getmantsev, G. G., and V. O. Rapoport (1960), On the build-up of electromagnetic waves in a plasma moving in a non-dispersive dielectric in the presence of a constant magnetic field, Sov. Phys. JETP, 11, 871-875.
539. Gold, L. (1966), Stream instabilities in magneto-ionic plasmas, Proceedings of the Seventh International Conference on Phenomena in Ionized Gases, vol. II, Beograd (Gradevinska Enjiga Publ. House, Belgrad, Yugoslavia).
540. Imre, K. (1962), Oscillations in a relativistic plasma, Phys. Fluids, 5, 459-466.
541. Landecker, K. (1952), Possibility of frequency multiplication and wave amplification by means of some relativistic effects, Phys. Rev., 86, 852-855.
542. Lichtenberg, A. J., and J. S. Jayson (1965), Propagation and instabilities of waves in bounded finite temperature plasmas, J. Appl. Phys., 36, 449-455.

543. Maxum, B. J., and A. W. Trivelpiece (1965), Two-stream cyclotron and plasma wave interactions, J. Appl. Phys., 36, 481-494.
544. Papa, R. J. (1965), The nonlinear interaction of an electromagnetic wave with a time-dependent plasma medium, Canad. J. Phys., 43, 38-56.
545. Romanov, Y. A., and G. F. Fillippov (1961), The interaction of fast electron beams with longitudinal plasma waves, Sov. Phys. JETP, 13, 87.
546. Rukhadze, A. A. (1962), Interaction of charged particle beam with a plasma, Sov. Phys. Tech. Phys., 7, 488.
547. Sagdeev, R. Z., and V. D. Shafranov (1961), On the instability of a plasma with an anisotropic distribution of velocities in a magnetic field, Sov. Phys. JETP, 12, 130-132.
548. Shapiro, V. D., and V. I. Shevchenko (1962), The non linear theory of interaction between charged particle beams and a plasma in a magnetic field, Sov. Phys. JETP, 15, 1053.
549. Storey, L. R. O. (1953), An investigation of whistling atmospherics, Phil. Trans., A, 246, 113-141.
550. Sturrock, P. A. (1958), Kinematics of growing waves, Phys. Rev., 112, 1488-1503.
551. Tsytovich, V. N. (1961), Spatial dispersion in a relativistic plasma, Sov. Phys. JETP, 13, 1249.
552. Twiss, R. Q. (1950), On Bailey's theory of growing circularly polarized waves in a sunspot, Phys. Rev., 80, 767-768.
553. Twiss, R. Q. (1951), On Bailey's theory of amplified circularly polarized waves in an ionized medium, Phys. Rev., 84, 448-457.
554. Vlaardingerbroek, M. T., and K. R. U. Weimer (1963), On wave propagation in beam-plasma systems in a finite magnetic field, Philips Res. Rep., 18, 95-108.

K. RELATED PUBLICATIONS BY THE AUTHORS

555. Bell, T., R. Smith, N. Brice, and H. Unz (1963), Comments on the magneto-ionic theory for a drifting plasma, IEEE Trans., AP-11, 194-195.
556. Chawla, B. R., D. Kalluri, and H. Unz (1967), Propagation of oblique electromagnetic waves through a warm plasma slab with normal magnetostatic field, Radio Science, 2, 869-879.

557. Chawla, B. R., S. S. Rao, and H. Unz (1966), Wave propagation in homogeneous gyrotropic compressible drifting plasma, J. Appl. Phys., 37, 3563-3566.

558. Chawla, B. R., S. S. Rao, H. Unz, D. K. Cheng, and H. C. Chen (1966), On the constitutive parameters in compressible gyrotropic media, IEEE Proc., 54, 812-814.

559. Chawla, B. R., and H. Unz (1966), On oblique incidence in a drifting anisotropic plasma, IEEE Proc., 54, 332.

560. Chawla, B. R., and H. Unz (1966), Potential equations in homogeneous isotropic moving media, IEEE Proc., 54, 397.

561. Chawla, B. R., and H. Unz (1966), On the refractive index equation for the magneto-ionic theory for drifting plasma, IEEE Trans., AP-14, 407-408.

562. Chawla, B. R., and H. Unz (1966), Radiation in a moving anisotropic plasma, IEEE Proc., 54, 1103-1105.

563. Chawla, B. R., and H. Unz (1966), On the magneto-electric effect, IEEE Proc., 54, 1102, 2033

564. Chawla, B. R., and H. Unz (1966), A note on the Lorentz transformations for a moving anisotropic plasma, Radio Science, 1, 1055.

565. Chawla, B. R., and H. Unz (1966), Relativistic magneto-ionic theory for drifting compressible plasma in longitudinal direction, IEEE Proc., 54, 1214-1215.

566. Chawla, B. R., and H. Unz (1967), Normal incidence on semi-infinite longitudinally drifting magneto-plasma: the relativistic solution, IEEE Trans., AP-15, 324-326.

567. Chawla, B. R., and H. Unz (1967), Wave propagation in a moving plasma, I.E.E. (London) - Electronics Letters, 3, 244-246.

568. Chawla, B. R., and H. Unz (1967), Poynting vector in moving plasmas, IEEE Proc., 55, 1741-1742.

569. Chawla, B. R., and H. Unz (1968), Reflection and transmission of normally incident waves by a semi-infinite longitudinally drifting magneto-plasma, Il Nuovo Cimento, 57-B, 399-418.

570. Chawla, B. R., and H. Unz (1969), Propagation in a two stream magneto-plasma, IEEE Trans., AP-17, 384-385.

571. Chawla, B. R., and H. Unz (1969), Reflection and transmission of electromagnetic waves normally incident on a plasma slab moving

uniformly along a magneto-static field, IEEE Trans., AP-17, 771-777.

572. Chawla, B. R., and H. Unz (1969), Wave propagation in relativistically moving warm magneto-plasma, IEEE Trans., AP-17, 822-823.

573. Compton, R. T., B. R. Chawla, and H. Unz (1966), Potential functions for moving media, IEEE Proc., 54, 980.

574. Epstein, M., and H. Unz (1963), Comments on two papers dealing with electromagnetic wave propagation in moving plasmas, IEEE Trans., AP-11, 193-194.

575. Kalluri, D., and H. Unz (1967), The first order coupled differential equations for waves in inhomogeneous warm magnetoplasmas, IEEE Proc., 55, 1620-1621.

576. Koyama, M., B. R. Chawla, and H. Unz (1967), On the polarization and the convection current models, IEEE Proc., 55, 579-581.

577. Rao, S. S., and H. Unz (1967), Asymptotic solution of oblique waves in inhomogeneous vertically magnetised plasma, IEE (London) - Electronics Lett., 3, 184-185.

578. Rao, S. S., and H. Unz (1968), Asymptotic solutions of waves in inhomogeneous plasma with longitudinal magnetic field, J. Geophys. Res., 73, 415-420.

579. Rao, S. S., and H. Unz (1968), On gradient coupling in inhomogeneous warm plasmas, Astrophys. J., 151, 797-798.

580. Sidhu, D. P., and H. Unz (1967), The magneto-ionic theory for drifting plasma-the whistler mode, Trans. Kansas Acad. Sci., 70, 432-450.

581. Unz, H. (1961), On vertical drift velocities of the F2 layer, J. Atmos. Terr. Phys., 21, 237-242.

582. Unz, H. (1961), Propagation in transversely magnetized plasma between two conducting parallel planes (includes bibliography), Antenna Lab., Ohio State Univ., Tech. Rep. 1021-9, Columbus, Ohio.

583. Unz, H. (1961), Electromagnetic waves in longitudinally magnetized ferrites between two conducting parallel planes, Antenna Lab., Ohio State Univ., Tech. Rep. 1021-10, Columbus, Ohio.

584. Unz, H. (1962), Propagation in arbitrarily magnetized ferrites between two conducting parallel planes (includes bibliography), Antenna Lab., Ohio State Univ., Tech. Rep. 1021-22, Columbus, Ohio.

585. Unz, H. (1962), The magneto-ionic theory for drifting plasma, IRE Trans., AP-10, 459-464.

586. Unz, H. (1963), Propagation in arbitrarily magnetized ferrites between two conducting parallel planes, IEEE Trans., MTT-11, 204-210.

587. Unz, H. (1963), The magneto-ionic theory for bound electrons, J. Atmos. Terr. Phys., 25, 281-286.

588. Unz, H. (1963), Electromagnetic radiation in drifting Tellegen anisotropic medium, IEEE Trans., AP-11, 573-578.

589. Unz, H. (1963), Oblique incidence on general magneto-plasma slab, Antenna Lab., Ohio State Univ., Tech. Rep. 1116-36, Columbus, Ohio.

590. Unz, H. (1965), Drifting plasma magneto-ionic theory for oblique incidence, IEEE Trans., AP-13, 595-600.

591. Unz, H. (1965), Oblique incidence on plane boundary between two general gyrotropic plasma media, J. Math. Phys., 6, 1813-1821.

592. Unz, H. (1966), On the derivation of Booker's quarti from Appleton-Hartree equation, IEEE Proc., 54, 304.

593. Unz, H. (1966), Wave propagation in drifting isotropic warm plasma, Radio Science, 1, 325-338.

594. Unz, H. (1966), The coupled wave equations for homogeneous gyrotropic compressible warm plasmas, IEEE Proc., 54, 393-394.

595. Unz, H. (1966), Asymptotic solution for inhomogeneous media, IEEE Proc., 54, 674.

596. Unz, H. (1966), Relativistic magneto-ionic theory for drifting plasma in longitudinal direction, Phys. Rev., 146, 92-95.

597. Unz, H. (1966), Oblique wave propagation in a compressible general magneto-plasma, Appl. Sci. Res., 16, 105-120.

598. Unz, H. (1966), On wave propagation in isotropic warm plasmas, IEEE Trans., Education, E-9, 149-154.

599. Unz, H. (1966), Direct integration of time dependent Maxwell equations, Am. J. Phys., 34, 1015-1019.

600. Unz. H. (1967), Asymptotic solution of waves in inhomogeneous plasma, IEEE Trans., AP-15, 494-495.

601. Unz, H. (1967), Wave propagation in inhomogeneous gyrotropic warm plasmas, Am. J. Phys., 35, 505-508.

602. Unz, H. (1967), Normal incidence on semi-infinite longitudinally drifting magneto-plasma: The nonrelativistic solution, IEEE Trans., MTT-15, 432-433.

603. Unz, H. (1967), Radiation from a sphere in compressible plasma, Appl. Sci. Res., 17, 340-354.

604. Unz, H. (1967), Oblique electromagnetic radiation incident on a semi-infinite warm general magneto-plasma, Il Nuovo Cimento, B-50, 207-223.

605. Unz, H. (1968), Relativistic magneto-ionic theory for drifting plasma, Radio Science, 3, 295-298.

606. Unz. H. (1968), Oblique wave propagation in drifting isotropic warm plasma, Radio Science, 3, 305-306.

607. Unz, H. (1968), Radiation from electromagnetic sources in an unbounded warm magneto-plasma, Il Nuovo Cimento, B-55, 481-490.

608. Unz, H., and R. S. Elliott (1967), Relativity and electricity, IEEE Spectrum, 4, 120-122.

609. Unz, H., and H. N. Kritikos (1967), Comments on the eikonal equation in a moving medium, IEEE Proc., 55, 1768.

610. Unz, H., and H. Zorski (1967), On the analogy between linear isotropic plasma theory and classical elasticity, Bull. Polish Acad. Sci., Warsaw, Poland, 15, 543-550.

ABOUT THE AUTHORS

Basant R. Chawla was born on January 30, 1944. He received the B.E. (honors) degree from the University of Bombay, India in 1963; the M.S.E. degree from Princeton University, Princeton, New Jersey, in 1965; and the Ph.D. (honors) degree from the University of Kansas, Lawrence, Kansas in 1967. Since 1967 he has been a member of the Technical Staff, Bell Telephone Laboratories, Inc., Murray Hill, New Jersey. He has done research work in control systems, wave propagation in plasmas, and instabilities in semiconductors, and he has more than twenty scientific publications.

Hillel Unz was born on August 15, 1929. He volunteered for the Palmach (Commando) and served in the Israel Defence Forces during the War of Independence, 1947-1949. He received the B.S. (Summa Cum Laude) degree in 1953 and the Diplome Ingenieur degree in 1961 from Israel Institute of Technology (Technion), Haifa, Israel; he received the M.S. degree in 1954 and the Ph.D. degree in 1957 from the University of California, Berkeley, California. Since 1957 he has been a faculty member in the Electrical Engineering Department, University of Kansas, Lawrence, Kansas, where he became a Professor in 1962.

During 1963-64 Dr. Unz was an NSF Postdoctoral Fellow in the Cavendish Laboratory, Cambridge University, England, and UNESCO Visiting Professor at the Mathematics Department, Israel Institute of Technology, Haifa, Israel. He has been a consultant to numerous research organizations, including the RAND Corporation, Santa Monica, California. He has done research work in electromagnetic theory, antennas, ionospheric and plasma propagation, and he has more than one hundred scientific publications. Dr. Unz was the first (1956) to suggest the use of the non-uniformly spaced antenna arrays.

SUBJECT INDEX

AUTHOR INDEX